Generic Intelligent Driver Support

Generic Intelligent Driver Support

Edited by

John A. Michon

Traffic Research Centre, University of Groningen, The Netherlands

Taylor & Francis
London • Washington, DC

UK   Taylor & Francis Ltd., 4 John Street, London WC1N 2ET

USA   Taylor & Francis Inc., 1900 Frost Road, Suite 101, Bristol, PA 19007

British Library Cataloging in Publication Data
A catalogue record for this book is available from the British Library

ISBN 0 7484 0069 9

Library of Congress Cataloging in Publication Data are availabe

Cover design by Amanda Barragry

Printed in Great Britain by Burgess Science Press, Basingstoke on paper which has a specified pH value on final paper manufacture of not less than 7.5 and therefor 'acid free'.

Contents

Project Colophon ix

List of Contributors xi
Preface xiii

Part I: The GIDS concept 1

Chapter 1: Introduction: a guide to GIDS 3
    John A. Michon, Alison Smiley
    1.0 Chapter outline 3
    1.1 What is GIDS? 3
    1.2 Practical objectives and core activities 6
    1.3 Origin and course of the project 8
    1.4 GIDS in the broader RTI context 8
    1.5 The domain of GIDS 10
    1.6 The functional characteristics of GIDS 12
    1.7 The GIDS implementations 15
    1.8 The scope of this book 17
    References 17

Chapter 2: Driving: task and environment 19
    Hans Godthelp, Bertold Färber, John A. Groeger, Guy Labiale
    2.0 Chapter outline 19
    2.1 The environment-vehicle-driver system 19
    2.2 Driving task analysis 21
    2.3 Infrastructure 24
    2.4 Vehicle 25
    2.5 Road transport informatics 28
    2.6 Conclusions 30
    References 30

Chapter 3: The driver 33
  Talib Rothengatter, Håkan Alm, Marja J. Kuiken, John A. Michon,
  Willem B. Verwey
  3.0 Chapter outline 33
  3.1 The analysis of driver characteristics 33
  3.2 Analyzing the driver's task...39
  3.3 Driver information needs..40
  3.4 Driver workload...42
  3.5 Conclusions 46
  References 47

Chapter 4: Driver support...53
  Wiel H. Janssen, Håkan Alm, John A. Michon, Alison Smiley
  4.0 Chapter outline 53
  4.1 The concept of driver support 53
  4.2 Navigation 61
  4.3 Manoeuvring: collision avoidance 62
  4.4 Control 63
  4.5 Behavioural feedback and instruction 64
  4.6 Collateral behaviour (multitasking) 65
  4.7 Conclusion..65
  References 66

Part II: The GIDS system 67

Chapter 5: Design considerations 69
  John A. Michon, Ep H. Piersma, Alison Smiley, Willem B. Verwey,
  Eamonn Webster
  5.0 Chapter outline 69
  5.1 The GIDS design philosophy 69
  5.2 Constraints...75
  5.3 Implementation...81
  5.4 Adaptivity in perspective..84
  5.5 Conclusion 86
  References 86

Chapter 6: GIDS intelligence 89
  Henry B. McLoughlin, John A. Michon, Wim van Winsum, Eamonn Webster
  6.0 Chapter outline 89
  6.1 Introduction 89
  6.2 The need for intelligence in a vehicle 90

6.3 Modelling the driver 91
6.4 Task analysis 94
6.5 Situation analysis 101
6.6 Architecture of the system 105
6.7 Conclusion: the generic aspects 110
References 111

Chapter 7: GIDS functions 113
Willem B. Verwey, Håkan Alm, John A. Groeger, Wiel H. Janssen,
Marja J. Kuiken, Jan Maarten Schraagen, Josef Schumann,
Wim van Winsum, Heinz Wontorra
7.0 Chapter outline 113
7.1 Navigation support 113
7.2 Collision avoidance support 119
7.3 Active control devices 123
7.4 Adaptive support 129
7.5 GIDS calibrator: individual preferences 136
7.6 Integrating functions in the GIDS system 140
References 142

Chapter 8: GIDS Architecture 147
Ep H. Piersma, Sjouke Burry, Willem B. Verwey, Wim van Winsum
8.0 Chapter Outline 147
8.1 Introduction 147
8.2 Connection to the outside world: sensors & navigation system 151
8.3 Supporting the driving task: Analyst/Planner 155
8.4 Supporting and improving the driver qua driver: PSALM 155
8.5 The Dialogue Controller 156
8.6 Control support: active car controls 163
8.7 The GIDS sytem hardware 163
8.8 Implementing GIDS in a vehicle 1: The TRC simulator 168
8.9 Implementing GIDS in a vehicle 2: The TNO-IZF Instrumented Car
for Computer Assisted Driving (ICACAD) 168
8.10 Conclusion 172
References 172

Chapter 9: GIDS Small World simulation 175
Wim van Winsum, Peter C. van Wolffelaar
9.0 Chapter outline 175
9.1 Functionality of the Small World simulation 175
9.2 The structure of the simulator 177
9.3 The traffic environment 181

9.4 Utilities 184
9.5 The Small World simulation as a test environment 186
9.6 Experimental issues and data collection 190
References 191

Part III: Performance and perspective..193

Chapter 10: Evaluation studies..195
Wiel H. Janssen, Marja J. Kuiken
10.0 Chapter outline 195
10.1 Introduction to the behavioural experiments 195
10.2 Small World simulation study...198
10.3 Field evaluation 206
10.4 General conclusions: what about GIDS? 213
References 214

Chapter 11: Impact and acceptance 217
John A. Groeger, Håkan Alm, Rudi Haller, John A. Michon
11.0 Chapter outline 217
11.1 Marketing GIDS 217
11.2 Acceptance of GIDS systems...220
11.3 Impact of GIDS 223
11.4 GIDS in a wider context..225
References 227

Chapter 12: The next steps...229
John A. Michon
12.0 Chapter outline 229
12.1 Have the GIDS objectives been achieved? 229
12.2 The further evolution of GIDS 232
12.3 Epilogue...236

Author index..241

Subject index 245

GIDS Project Colophon

Generic Intelligent Driver Support (GIDS) has been studied as project V1041 of the DRIVE programme of the European Economic Community. It brought together 11 partners from 5 EEC countries and 2 partners from the EFTA country Sweden. The GIDS Consortium was composed in the following way.

Partners
  BMW AG (D)
  Delft University of Technology (NL)
  INRETS-LEN (F)
  MRC-Applied Psychology Unit (GB)
  Philips Research Laboratories (NL)
  Saab-Scania AB (S)
  Swedish Road and Traffic Research Institute VTI (S)
  TNO-Institute for Perception (NL)
  Traffic Research Centre, University of Groningen (NL)
  Trégie Groupe Renault (F)
  University College Dublin (IRL)
  University of the Armed Forces Munich (D)
  Yard Ltd (GB)

Project Coordination
  Prof.Dr. John A. Michon (Project Coordinator)
  Drs. Marja J. Kuiken (Deputy Project Coordinator)
  Traffic Research Centre, University of Groningen (NL)

List of Contributors

ALM, Håkan, Swedish Road and Traffic Research Institute VTI, Linköping, Sweden

BROWN, Ivan D., MRC-Applied Psychology Unit, Cambridge, United Kingdom

BURRY, Sjouke, TNO-Institute for Perception, Soesterberg, The Netherlands

FARBER, Berthold, University of the Armed Forces Munich, Munich, Germany

GODTHELP, Hans, TNO-Institute for Perception, Soesterberg, The Netherlands

GROEGER, John A., MRC-Applied Psychology Unit, Cambridge, United Kingdom

HALLER, Rudi, BMW-AG, Munich, Germany

JANSSEN, Wiel H., TNO-Institute for Perception, Soesterberg, The Netherlands

KUIKEN, Marja J., Traffic Research Centre, University of Groningen, Haren, The Netherlands

LABIALE, Guy, INRETS-LEN, Bron, France

MCLOUGHLIN, Henry B., University College Dublin, Dublin, Ireland

MICHON, John A., Traffic Research Centre, University of Groningen, Haren, The Netherlands

MILTENBURG, Pieter G.M., Traffic Research Centre, University of Groningen, Haren, The Netherlands

PIERSMA, Ep H., Traffic Research Centre, University of Groningen, Haren, The Netherlands

ROTHENGATTER, Talib, Traffic Research Centre, University of Groningen, Haren, The Netherlands

SCHRAAGEN, Jan Maarten, TNO-Institute for Perception, Soesterberg, The Netherlands

SCHUMANN, Josef, University of the Armed Forces Munich, Munich, Germany

SMILEY, Alison, Human Factors North Inc., Toronto, Canada

VERWEY, Willem B., TNO-Institute for Perception, Soesterberg, The Netherlands

WEBSTER, Eamonn, University College Dublin, Dublin, Ireland

WINSUM, Wim van, Traffic Research Centre, University of Groningen, Haren, The Netherlands

WOLFFELAAR, Peter C. van, Traffic Research Centre, University of Groningen, Haren, The Netherlands

WONTORRA, Heinz, University of the Armed Forces Munich, Munich, Germany

Preface

The area of Road Transport Informatics (RTI) is evolving rapidly. This lends some urgency to the question of the extent to which human operators will still be able to perform the task of conducting their vehicle efficiently, effectively and safely if they are facing a driving environment that is increasingly dominated by automated and semi-automated information and control systems. It is with this question in mind that we have undertaken to develop the Generic Intelligent Driver Support system known, for short, as *GIDS*. GIDS helps the driver to cope with the avalanche of information that we may expect to be overwhelming road traffic in the next decade or so.

Why, one may ask, have we focused in this project on driver support – signalling, warning, and advising – rather than on a system that may take over control from the driver altogether? Why didn't we make a more principled step in the direction of a robot driver as I had suggested earlier (Michon, 1987)[1]. The answer is simple. In order to introduce cooperative or automatic driving, a traffic environment will be needed that is totally different from the present, comparatively unconstrained environment which admits vehicles, road technologies and drivers, each and all with vastly different performance characteristics. Until major steps towards a much more tightly constrained road environment come within view, it would seem advisable to maintain and even increase the effort devoted to the kind of driver support envisioned in GIDS and to adopt an appropriately cautious attitude towards cooperative or automated driving. For the time being, however, nothing even remotely approaching a sufficiently constrained environment appears to be around.

The GIDS Project has been a rather complicated exercise in human-vehicle environment interfacing. As such it has achieved its principal goals: we have developed a GIDS prototype, a truly state-of-the-art intelligent driver support system, performing a non trivial subset of the support functions such a system should eventually incorporate. In this volume the partners who participated in the

---

1   Michon, J.A. (2 September, 1987). Twenty-five years of road safety research. Opening Adress to the Second International Road Safety Conference, Groningen, The Netherlands, 1-4 September, 1987. *Staatscourant, no. 168,* 4-6.

GIDS Project present an account of the considerations and activities that carried us so far.

What follows, however, is not just a description of the GIDS prototypes now available. It is also an account of a fairly major international co-operative Research & Development venture on a European scale. Altogether 13 partners from six European countries have contributed to this development. Looking at the current problems with the European integration, the GIDS Consortium should be allowed to take some pride in its achievement. Also, I know of few projects in the field of road traffic and transportation that have been so critically dependent on the collaboration of engineers and behavioural scientists. And last but not least, I know of no other project where behavioural scientists eventually took the responsibility for the effective realization of a technical system.

Although the objectives to be met seemed clear, right from the beginning, the succession of steps to be taken was not quite so evident. Several sudden, unexpected and almost serendipitous events happened along the route followed.

A first illustration coming to mind is the initial difficulty of some of the engineers involved in the project to understand the implications of a dynamic information processing model put forward by the experimental psychologists in the consortium. The psychologists argued in vain that the model they were proposing was just a special case of the more general and flexible knowledge processing model until, quite by chance, the name Seeheim was introduced into the discussion. Apparently our engineers stood in awe of the Seeheim model – a name unfamiliar to psychologists. When it transpired that the psychologists' model was, in fact, formally equivalent with the Seeheim model a torrent of mutual understanding broke loose.

As a second illustration I may refer to the difficult methodological start we experienced. One of the first stated aims was to develop a strategy allowing the behavioural scientists and the software developers to communicate. Initially we had in mind to decide on a common set of analytic techniques and experimental procedures. Soon, however, we found out that this was going to impose too many conceptual restrictions on partners' individual contributions. But then, after nearly two days, of heated but hardly productive discussion, suddenly the idea of the *Small World* paradigm emerged, just as the goddess Athena sprang from Zeus' head, full-blown and in battle armour: the Small World proved instantly to be the proper creative framework for many of our methodological and conceptual problems.

It is events such as these that make participation in projects such as GIDS so eminently rewarding: they occur usually at unguarded moments of mild despair and almost always after considerable intellectual effort on the part of those involved. It is especially for the shared joy of such events – and there were quite a number in the course of the project – that I wish to thank my partners. Although, as Project Coordinator, I occasionally felt like a gardener pushing a wheelbarrow-full of lively frogs, I wish to emphasize that this is just the way creative colleagues should behave, whether I like it or not. This is, incidentally, not entirely what the

DRIVE Central Office in Brussels appeared to believe and if, during the project, I have failed in some respect, it was that I have not been able to keep the bureaucracy sufficiently transparent for my partners.

By any standard the GIDS Project has been extremely rich in output. Not only do we now have two operational systems, one in an automobile for real-world studies and an identical one in an advanced driving simulator, but in addition there are well over 30 reports to the European Commission. Moreover there is a variety of contributions to scientific media, articles in reviewed journals, and chapters in books and conference proceedings. Their number is still growing and this may be expected to continue for some time to come, especially in view of the ARIADNE Project (DRIVE II Project V2004) which continues the development of GIDS.

This book summarizes what the GIDS Consortium has achieved in the course of the past three years (and a half). But rather than publishing an anthology of individual papers we decided to produce a more rigorously edited volume with one person bearing the editorial responsibility for the entire text, but with the assistance of a chapter editor for each of the 12 chapters. These chapter editors were, in their turn, supplied by the individual investigators with a variety of shorter and longer textual fragments. Although the individual contributions are no longer recognizable, the contributors to each chapter have been acknowledged as such at the beginning of each chapter: the contributing authors are mentioned in alphabetical order, following the chapter editor's name.

As general editor I wish to thank the chapter editors and all contributors for their concerted efforts to supply me with the necessary building bricks. When, 20 years ago, I was editing the Dutch *Handboek der Psychonomie,* I first formulated my Law of Editorial Misery. It states that the burden on an editor increases proportionally with the square of the number of contributors. With the present volume completed I now realize that it is not so bad. I now have reasons to believe that there is a ceiling effect, so that the total burden will eventually reach an upper limit... unless the contributors to this volume have been unusually cooperative. Perhaps they know better than I, but I seem to have some reason for satisfaction even if as a result the Law of Editorial Misery must be revised.

At this point I might personally thank all those who contributed in one way or another to the successful completion of the GIDS Project. After some deliberation, however, I found this totally impossible. Instead I simply conclude that nearly everyone involved in the GIDS Project has contributed to the ultimate product in an essential way. Yet, as Project Coordinator and as Editor I wish to single out five individual members of the GIDS project crew at the Traffic Research Centre.

Running a project the size of GIDS requires considerable managerial back up. The conditions of the DRIVE contract and my relentlessly continuing responsibilities at the University of Groningen's Psychology Department bore heavily on Marja Kuiken, Talib Rothengatter and Suzanne Punt.

Marja Kuiken has been a wonderful Deputy Project Coordinator. Always pleasantly cooperative, and always *ad rem* so that I could immediately tell if things were going smoothly or not, as soon as we would get together. If not, as would occasionally be the case, she would usually have initiated the appropriate action to keep things under control or she would at least suggest a plausible course of action.

Apart from being involved in GIDS directly as an investigator, Professor Talib Rothengatter also played an important role in his capacity as co-director of the Traffic Research Centre, facilitating in many ways the managerial efforts to keep GIDS on course, especially the financial issues that consistently failed to arouse my interest. But even more importantly Talib was, right from the beginning of the DRIVE Programme, "our man in Brussels". He became thoroughly involved in the extensive attempts at coordinating various DRIVE projects, acting as chairman of various committees and study groups, and carrying through much of the negotiations with DRIVE Central Office. This he did in a way that has been very important for the overall status and credibility of GIDS in the DRIVE 'scene'.

In the course of the project Suzanne Punt evolved into a perfect office manager, acquiring a subtly insistent style of handling 1001 and more things with DRIVE Central Office and with the 13 partners. Not only did she take care of administering and delivering the required monthly, quarterly, and annual reports, the provisional and definitive contracts and cost statements, plus all first, second and higher revisions and amendments to the aforementioned provisional and definitive contracts and cost statements. She also managed to persuade the GIDS partners to provide the required manuscripts and other materials for deliverables and chapters, with revisions and revisions of revisions from partners, transforming them gradually into bodies of text that I gradually began to recognize as suitable chapters for this volume. Don't ask me how she did it, but she did it. And all partners love her for it. And so does the editor!

At this point I wish to mention Professor Ivan Brown who has kindly shared part of my responsibility as an editor by looking over my shoulder and ploughing through the entire manuscript in two of its earlier versions, wielding his linguistic axe and also using his fine sense for semantic and pragmatic coherence of the text. With a consortium incorporating no less than six native tongues one should not expect the Nobel prize for literature, but in the unlikely event that one day we will be nominated it will to a large extent be thanks to Ivan Brown. It helped a great deal that Professor Brown held the extramural Chair of Traffic Science at the University of Groningen during the project.

Finally, having been involved in a good number of complicated research projects, I know that in any such project there is at least one point at which the whole thing seems to be on the verge of collapsing under its own weight. When you are building a Triumphal Arch this crucial moment proverbially happens when you are trying to put the keystone in place. For GIDS the critical moment arrived when Philips Research Laboratories had to withdraw from the project as a result of a major corporate decision, early in 1991. We were, then, exactly two years into the project and were just completing the system specifications, more or less according to plan. This withdrawal left the GIDS Consortium essentially without its electronic engineers. At that point the GIDS Project would have collapsed had not Ep Piersma, a psychologist by training and vocation, offered himself voluntarily as chief engineer of the project to undertake the major task of composing the actual GIDS hardware from whatever commercially available units he could lay his hands on. This involved, among other things, writing specifications for components; contracting with a dozen or so small and not so small companies; bringing together the resulting bits and pieces; praying that together they would still function according to specification and modifying them when his prayers were not answered. The reader should realize that from an engineering point of view this approach is unconventional, to say the least. Undisturbed by the raised eyebrows and other telltale signs of disbelief in his immediate environment, Ep pressed on (with the indispensable outside help of Ronald Vossen of BSO). We now know Ep was right and all those who have been involved in the GIDS Project should be grateful for his courage and his exceptional vision.

John A. Michon
GIDS Project Coordinator
Editor

Haren, The Netherlands
25 December 1992

Part I
The GIDS concept

Chapter 1
Introduction: a guide to GIDS

John A. Michon, Alison Smiley

1.0 Chapter outline

In this introductory chapter an overview is presented of the nature and development of GIDS, the *Generic Intelligent Driver Support* system. The development of GIDS took place between 1989 and the middle of 1992, as part of the EEC DRIVE programme. The present volume, summarizing the activities of the GIDS Consortium, was prepared at the conclusion of the project.

An overview of an intelligent, or adaptive, driver support system is presented in Section 1.1. Next the objectives (Section 1.2) and some historical background (Section 1.3) of the GIDS project are provided. The position of GIDS within the larger framework of Road Transport Informatics (RTI) is reviewed in Section 1.4.

The next four sections present the GIDS perspective on driver support. In Section 1.5 we outline the domain that is covered by the GIDS concept and the constraints that had to be imposed to keep the project within practical limits. In Section 1.6 we review the functional characteristics of the GIDS system, and Section 1.7 is devoted to the way in which GIDS has been implemented.

Finally, in Section 1.8 a brief preview of the subsequent chapters of this book is presented.

1.1 What is GIDS?

Introduction

The overall objective of the GIDS project has been "to determine the requirements and design standards for a class of intelligent co-driver (GIDS) systems that are maximally consistent with the information requirements and performance capabilities of the human driver" (from the GIDS project proposal, GIDS, 1988). The project has resulted in recommendations for such systems and in a (limited) operational prototype demonstrating the essential features of the GIDS concept, both under simulated and real world traffic conditions. The GIDS project – officially known as DRIVE Project V1041 "Generic Intelligent Driver Support (GIDS)" – was part of the DRIVE Programme which was initiated in 1988 by the Commission of the European Communities to stimulate and coordinate the introduction of modern Road Transport Informatics (RTI). DRIVE stands for Dedicated Road Infrastructure for Vehicle safety in Europe.

Co-driver systems or driver support systems – the latter being the term to be used throughout this book – derive their usefulness to a large extent from the fact that vehicle operators must cope with a growing amount of information of an increasingly complex nature. This is caused by several factors, including increasing traffic density, an increasing number of on-board and roadside sources of information and, last but not least, by the increasing amount of additional in-vehicle equipment, such as telephones and fax machines. A driver support system, such as GIDS, will help to counter the information pollution that is threatening the vehicle operator, by filtering, interpreting, integrating, prioritizing, and presenting the information from any number of sensors and applications.

This avalanche of information – much of which will eventually be generated by RTI systems resulting from the DRIVE programme – is likely to have an impact on almost every aspect of the driving task. It will affect route planning as well as navigation, manoeuvring, and elementary vehicle control. Unless regulatory action is taken this information will eventually be presented to the driver in an essentially incoherent fashion, irrespective of its importance or appropriateness. A critical function of GIDS, as of any other driver support system, is to protect the driver from being overwhelmed by such uncoordinated information.

The innovative feature of GIDS is that it is the first system ever to take into account, in an adaptive fashion, (some of) the intentions, capabilities, and limitations of the individual driver. Driver support systems should enable drivers to cope with the driving task more easily, safely and efficiently (and, indirectly, at diminished cost to the environment).

By meeting its overall objective, the GIDS project will help to further the goals of the DRIVE programme which were: "to increase traffic safety, to improve transport efficiency, to reduce energy consumption and to improve the environment" (as stated in the DRIVE Call for Proposals, DRIVE, 1988).

A scenario

To obtain a feeling for what GIDS might eventually become, imagine you are travelling to Switzerland for a holiday in the Alps. Upon arrival your rental car is waiting for you at the airport. In order to get going, you simply insert your personalized GIDS smart card in the receptacle on the dashboard. Instantly the vehicle will recognize who you are and adapt automatically to your individual needs and traits. It will know, say, that you have never driven on mountain roads before, and that you have a tendency to brake very briskly. Knowing these and other things, the GIDS system will be prepared to guide you slowly through hairpins and keep you far from those nasty 15^+ per cent descending stretches of mountain road, at least for the first couple of days. Moreover, realizing that your knowledge of, say, French and Italian is marginal, it will translate all the local traffic information into your native language, including whatever there is to read on the various road signs.

The G, I, D, and S of GIDS

Whilst admitting that this is a somewhat far-fetched sketch of the performance of GIDS, let us now consider what GIDS is by looking more closely at the constituent concepts of the term.

The term *generic* refers to the fact that GIDS has been conceptualized in such a way that it may increase in scope and complexity with the development of new technological capabilities. The system as it figures in this volume covers a limited but realistic set of tasks, namely such tasks as can presently be defined unambiguously and for which equipment is currently feasible. GIDS is based on a communication protocol, that is, an agreed way of communication between the driver and the system. As a result the GIDS system and the external inputs to the system can be extended to accommodate subtasks and information sources that may be added later. In other words, GIDS is generic in the sense that it is not specific to the automation of particular driving tasks or rigidly defined categories of information.

GIDS contains *intelligence* of a kind that other current support systems do not. Ideally a GIDS system should be able to manage the stream of information to the driver in accordance with the driving task, the driving conditions, and the state the driver is in. Current support systems lack this type of intelligence. They do not select or integrate information according to the demands of the driving task or according to the state and intentions of the driver. Not only do conventional systems not select the appropriate level of detail, they also do not prevent the driver from selecting a level of detail that is inappropriate in the face of the current situation. With currently available technology drivers can be using a navigation system, talking on a cellular phone, and checking on their moment-to-moment fuel economy. The information from these multiple sources is not integrated in conventional systems and therefore unsafe situations may arise, due to overload or contradictory advice. GIDS is designed to prevent drivers from overloading themselves because it will integrate the available information from various sources, taking driver needs and intentions into account.

In designing GIDS to support a variety of driving tasks, the physical, perceptual and cognitive characteristics of the *driver* needed consideration. This includes his or her physical capacities, perceptual capacities, and cognitive capacities. An important additional characteristic of the driver which, nevertheless, has not been explicitly considered in the GIDS project, is the driver's emotional state. This deliberate omission has been an occasional source of criticism. After all, emotion is a powerful determinant of driving behaviour: a driver who is preoccupied with some stressful personal situation may be inattentive, and a driver who is feeling aggressive may speed, or overtake unsafely, and so on. However, only those aspects of emotional and motivational states that are observable, such as higher speeds, may be considered in a GIDS system, but thus far these aspects do not, unfortunately, provide unambiguous cues about the driver's state.

GIDS provides *support* at three major levels defining the driving task: navigation, manoeuvring and control. Besides support at each of these levels, the GIDS system provides support on a "meta-level". This is the area in which the GIDS concept differs most from conventional support systems.

First, information from individual support systems serving each of the three task levels is integrated and prioritized so that the system can respond appropriately to the driver's current situation. This means that the mental workload of the driver is taken into account in the presentation of information.

Second, the GIDS system can provide adaptive feedback. This includes evaluation of prior performance, instructive feedback, and progressive modification of support structure. The latter refers to such possibilities as the system gradually giving less and less route information as the driver has more experience of a particular route, down to the point where only information about changed conditions, such as a new construction zone, is presented.

1.2 Practical objectives and core activities

The GIDS overall objective, stated in Section 1.1, is too abstract to offer much practical guidance. In more concrete terms, the GIDS consortium has pursued the following set of practical goals, as stated in the original GIDS proposal:

- define detailed functional requirements of generic intelligent driver support (GIDS) systems;
- determine the impact of new road transport informatics (RTI) systems on the task representations and behaviours of drivers with respect to navigation, manoeuvring, and control aspects of the driving task;
- determine the interactive communication (display and dialogue) between the driver and the new RTI systems, inclusive of adaptive feedback;

- develop the required hardware and software that will lead to the implementation of a prototypical GIDS system. This prototype shall incorporate the substantive core of the GIDS concept;
- determine the impact of systems that meet the GIDS specifications on driving
  safety, efficiency, training, and system acceptance;
- demonstrate the validity of the GIDS concept in field tests.

These objectives have been pursued in two major stages. The first stage roughly
covered the first 18 months of the project. It dealt with the specification of behavioural requirements and technical characteristics for a GIDS prototype of limited
functionality. The second stage, initially intended to cover the period 1 July 1990
until 31 December 1991, but ultimately lasting until the middle of 1992, was
devoted to the implementation and evaluation of this prototype.

We distinguish four major component activities that, together, have enabled the
GIDS consortium to achieve the stated objectives according to plan.

The first activity consisted of a definition of the basic functional and operational
requirements of GIDS systems. For this purpose several basic functional domains
were distinguished. Altogether five such functional domains have been recognized
as basically covering the whole domain of driving tasks: planning, manoeuvring,
control, adaptive feedback instructions, and functional integration of non-driving
activities, such as carrying on a telephone conversation while driving. Within each
domain a characteristic application was then selected, allowing us to keep the size
and complexity of the prototype within reasonable bounds whilst retaining the essentially generic nature of the GIDS concept.

The second activity was a critical examination of the functional and operational
requirements of GIDS systems. For this purpose a series of preliminary studies –
literature reviews and pilot experiments – were carried out early in the project. The
results of these studies initially gave rise to a working definition of the GIDS concept and later to operational recommendations for GIDS systems in general and for
the GIDS prototype in particular.

The results of the component studies were integrated into concrete design specifications, incorporating the substantive core of GIDS system functions and operations. These specifications guided the construction of the prototype GIDS system.
Under the terms of the present project a prototype of limited functionality has
become available, in two versions. The first is implemented in a genuine automobile for on-the-road studies, the second is part of a simulation facility. Further
development is required before a reliable and practical driver support system that is
economically feasible can eventually emerge.

Once the prototypes became available, field tests were carried out to measure
the GIDS system's effectiveness and behavioural impact. This activity took up the
final part of the last project year. The evaluation proceeded on the basis of hardware and software performance tests and on the basis of a functional evaluation by

testing predictions about driver performance and acceptance, with special concern for the criteria specified, viz. safety, efficiency, and instruction. This stage finally converged on a field demonstration which effectively concluded the GIDS project.

1.3 Origin and course of the project

The ultimate aim of GIDS was to provide a driver with adaptive intelligent support. For this we must be able to produce a formal description of a driver's behaviour to the extent that it can be understood by an artificial intelligence, a computational system. The question is, of course, how long models that qualified for this purpose had been around before the inception of GIDS. The answer is: Not very long!

Only by 1984, it seems, had the intelligence of cognitive architectures become flexible enough to support an effective formalization of the driving task (Michon, 1985). This led to a semi-serious proposal to develop the intelligence required for a robot driver which (or should we say *who*?) would be able to pass its driver's licence examination by the year 2000 (Michon, 1987). At the time this was a somewhat exalted and far-fetched idea which met with a good deal of disbelief.

Then an excellent opportunity arose to test the feasibility of this idea on a more modest and more realistic scale, when the Commission of the European Communities launched the DRIVE programme. Following a series of confusing consultations, a brainstorming session was held in the summer of 1988, involving representatives of the Traffic Research Centre RUG, the TNO Institute for Perception, and Philips Bedrijven NV. Eventually, early in October of the same year, a consortium of thirteen universities, industrial companies, and research institutes from six European countries combined forces and submitted a proposal to the DRIVE Commission (GIDS, 1988). The proposal was negotiated and accepted, and work began on 28 January 1989. The project was officially completed on 31 December 1991 although some of the work went on beyond this deadline. The final report was accepted by the European Commission on 17 October 1992.

The outline structure of the GIDS project as it has been carried out is schematically represented in Figure 1.1.

1.4 GIDS in the broader RTI context

Several aspects of the GIDS concept put the project in a central position in the wider context of RTI as studied under the DRIVE programme. The GIDS approach is generic in the sense that it is not, in principle, attached to domain-, task-, or equipment-specific requirements. This implies a kind of flexibility that should eventually facilitate its customization for a broad spectrum of dedicated RTI applications, both roadside- and vehicle-based.

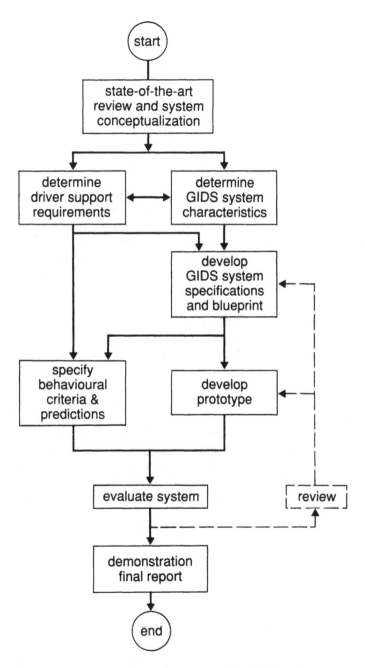

Figure 1.1 The main stages of the GIDS project

Thus, at a later stage, GIDS should be able to play a role in the development of 'smart roads' and 'smart cars' alike. The integrative nature of the GIDS concept should help to provide multifunctional support to drivers in a variety of situations and thus will help to avoid the much dreaded information explosion in road traffic that might otherwise overwhelm and confuse drivers, and influence their performance adversely.

At this point we wish to emphasize the complementarity of the GIDS project with other, related programs. The GIDS project is special in the sense that it provides an essential link for improving the important human-machine interface; pursuing a systems approach that takes full account of the human component in the system. Related activities in DRIVE and, for that matter, the PROMETHEUS programme, do not similarly proceed from this perspective. One aim of the GIDS project has been to impose realistic and psychologically meaningful constraints to ensure that these activities will be user-centered rather than technology-driven. Such meaningful constraints should not in any way interfere with informed and user-compatible technological or economic developments but, on the contrary, should support development by helping to avoid costly but unnecessary (useless or unacceptable) steps on the way to a large-scale implementation of road transport informatics.

The GIDS Project to be considered in this volume was subject to the inherent dynamics of the DRIVE programme. In this sense it has, undoubtedly, followed and even reinforced the trend towards the "informatization" of the traffic environment. It is appropriate at this point to sound two notes of warning. First, it should be emphasized that the introduction of GIDS systems has the potential to create a technology spiral. Effective driver support systems will enable the driver to cope with more and more complex information. This, in turn, may induce vehicle manufacturers, road administrators and the drivers themselves to expand and complicate the information delivered, which will call for more advanced driver support systems, and so on. Second, while GIDS systems certainly have the potential for achieving the DRIVE goals of increasing safety and efficiency of driving, through supporting drivers in particular tasks, there may be better ways of attaining the same goals. To take an example, it may be that building all the intelligence into individual systems in each car will be a more expensive and less effective means of maintaining lane position than building less intelligence into the car but adding to the road infrastructure instead, so that cars are automatically guided along high-speed roads.

1.5 The domain of GIDS

In GIDS, the question as to how intelligent driver support should be structured is stated in terms that are consistent with the present state of the art of road transport informatics and behavioural science methodology. The complex and underdetermined nature of the driving task makes it impossible to take the entire domain into account. A selection therefore had to be made. The proposed constraints allow for a limited but, nevertheless realistic and important set of features:

(a) Environmental conditions

The driving circumstances have been integrated into a subset of the real world – the so called Small World – which allows a driver to negotiate typical road environments, including a roundabout, T-junctions, an intersection and curved roads (Figure 1.2). Also a limited set of environmental conditions has been specified for this Small World: there may, for instance, be certain obstacles (vehicles, trees, buildings), and visibility may be either high or low. The Small World has guided much of the GIDS research. As such it has proved to be a powerful methodological heuristic. It has reduced the computational complexity of the GIDS system to acceptable proportions; and it can be implemented in the real world, as well as in driving simulator studies and computer simulations.

(b) Subtasks of the driving task

The tasks studied have been geared to the constraints imposed by the Small World and include entering and exiting a roundabout, turning, merging, negotiating an intersection, curve tracking, car following, and overtaking. Together these form a small but important set of (sub)tasks for which a definite, closed, computational description can be given and which cover a considerable percentage of ordinary driving actions.

(c) Support functions

The support functions realized in the GIDS prototype derive from the aim to provide driver support at each of the principal levels at which road users must cope with their task: planning (navigation), manoeuvring (obstacle avoidance), and handling (steering and accelerator control). In addition the role of instructional feedback to novice drivers and the effect of some concurrent in-vehicle tasks not directly related to driving (carrying on a telephone conversation) have been studied.

(d) System architecture

Functionally the GIDS architecture consists of an analyst/planner, accepting inputs from a series of special-purpose applications (sensors), a repertoire of 'situations', a data base containing information about the driver, and a dialogue controller. The hardware components of the system are integrated in a bus architecture allowing bidirectional communication between all components.

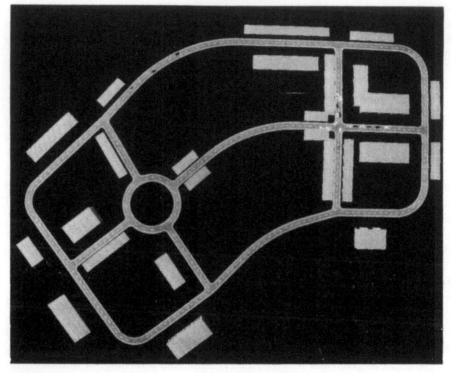

Figure 1.2 The Small World topography

(e) Presentation systems (human-machine interfacing)
The GIDS architecture allows drivers to interact with the GIDS system by means of a variety of displays and controls, including voice input and output, a keyboard and LCD display, switches, and active controls (steering wheel and accelerator).

(f) Driver characteristics
Finally the GIDS system is able to adapt to various states and traits of drivers. Initially a limited number of driver characteristics has been selected, in particular age and experience.

1.6 The functional characteristics of GIDS

The GIDS system operates in the following way. It obtains its inputs from any number of low-level sensors and intermediate-level (dedicated) support systems or applications, such as a navigation system or a collision avoidance system, each of which provides messages with its own format and its own domain-sensitive content or meaning. These front-end systems generate the GIDS system's knowledge about the world and thus provide the basis for the messages (warnings, advice, interventions) the system may subsequently decide to impart to the driver. Whether or not a particular message will be presented to the driver depends on the results of a comparison between the observed behaviour and the required behaviour, computed by the system, and on the driver's needs and intentions. It remains a matter for further study to determine to what extent and degree the GIDS architecture can explicitly and unambiguously infer these needs and intentions from its inputs: the limitations of the GIDS design cannot be defined in advance!

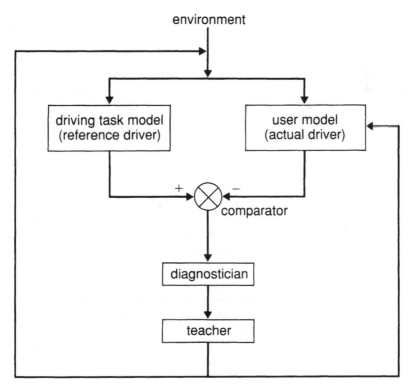

Figure 1.3 Conceptual model for GIDS

Once the driver's needs for support have been determined, the system will select from its knowledge repertoire a 'scenario' that represents a particular (sub)task to meet this requirement. As long as the behaviour of the driver stays within the accepted range, no message will filter through; otherwise the system will allow the relevant messages to be communicated to the driver. While this is the normal mode of operation for the GIDS system, it will, in the meantime, continuously monitor the actual situation and allow priority interrupts whenever necessary.

The model on which the GIDS system is based contains four primitive elements (Figure 1.3). In later chapters we will expand this elementary scheme into the actual GIDS system. For the present, however, understanding is best served by refering to these four primitives since they are related rather directly to similar components that make up 'intelligent tutoring systems' (e.g., Sleeman & Brown, 1982). Any adaptive support system must encompass these elements in one way or another.

(a) The reference driver model
The first component, the reference driver model, represents the normative state of affairs over the entire operational range for which it has been designed. It should, in other words, incorporate a number of features, including knowledge about the 'average' driver and the normal range of variations around the standard, legal restrictions, etc., that are required to perform the driving task – or at least certain specified parts of it – correctly. When supplied with sufficient and accurate information about the driving task and the driving environment (which includes the required information about the state of the vehicle), it will perform its task effectively, efficiently, and safely.

(b) The actual driver model
The actual driver model is a component incorporating the actual driver's representation, or 'knowledge' of the driving task and the traffic environment, including representations of systematic error forms; for instance, behavioural tendencies which are due to the driver's level of experience or capacity limitations.

(c) The diagnostician
The diagnostician will detect discrepancies between the behaviour of the actual driver and the reference driver. It contains knowledge allowing it to infer what ought to be done under the prevailing conditions and within the range of its available options. Eventually an intelligent support system must be able to build a record of previous behaviours and mismatches between required and actual performance of individual drivers and to infer from such a record what is causing any systematic error tendencies.

(d) The teacher
The didactic component decides how to communicate with the driver in terms of information channel, order and urgency of messages, and type of intervention. Here we have to deal with two aspects: first, immediate attempts at warning (e.g., to avoid an accident) and, second, long-term adaptation (within the system itself, including adaptations of the model of the actual driver).

1.7 The GIDS implementations

The GIDS automobile (ICACAD)

The GIDS project has been working towards an operational prototype implemented in an automobile (Figure 1.4). This vehicle, known as ICACAD, which stands for Instrumented CAr for Computer-Assisted Driving, is special in the sense that it not only incorporates a complete GIDS prototype system, but is also an experimental vehicle for behavioural studies. This implies that it is possible to measure behaviourally relevant variables of the vehicle, the environment, and the driver. This vehicle was specially developed at the TNO Institute for Perception within the framework of the GIDS project. It is based on an existing design of an instrumented vehicle for road user studies, modified so as to accommodate the GIDS prototype.

Figure 1.4 ICACAD, the GIDS demonstration vehicle

The Small World simulation

Despite the heavy constraints imposed on the overall system design by the Small World, it has been a major task to compile a sufficiently detailed inventory of driver performance data, to implement the various event representations, and to analyze the dialogue structure of the GIDS system. All this constitutes the required know-ledge base for a driver support system that we may, indeed, call intelligent. This re-quirement leads to the problem of creating and testing a sufficiently large set of event sequences, among other things to rule out the possibility of inappropriate warnings or instructions. Rather than realizing this complicated generate-and-test procedure from the armchair or in a costly real-world experimental programme, a Small World simulation was developed at the Traffic Research Centre of the University of Groningen (Figure 1.5).

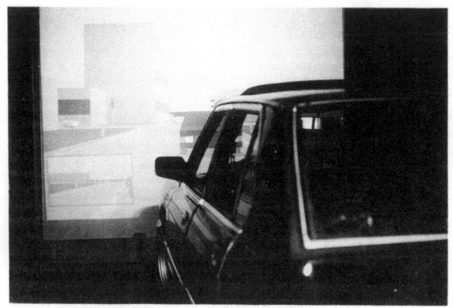

Figure 1.5 The TRC simulator, another GIDS demonstrator

This facility allows the testing of a wide range of event representations, message structures, and dialogue scheduling, that arise when driving through a simulated en-vironment. Thus it greatly expedites the process of implementing and extending the GIDS knowledge base, as this depends in large measure on our ability to identify an appropriate set of constraints on what otherwise would be an infinite set of possible, but largely dangerous, unacceptable actions. By selecting the proper con-straints, the most abstruse consequences of certain manoeuvres can be ruled out on an a priori basis. At the same time the Small World simulation allows an evaluation of the GIDS system prototype under simulated driving conditions, especially in those situations that are too hazardous for testing under real driving conditions.

1.8 The scope of this book

Part I of this book describes the scientific and technical foundation of the GIDS concept. This is based on the well-known subdivision of the road traffic system into three major components: road – vehicle – driver. In Chapter 2 the driving task and the driving environment are analyzed. In Chapter 3 the focus is on the characteristics of the driver. Then, in Chapter 4, we turn to the interactions between the driver, the vehicle and the environment and, more specifically, to the concept of driver support.

Part II is devoted to the actual GIDS system, its specification and implementation. First, in Chapter 5 the constraints imposed on the actual design are discussed in some detail. Some of these constraints were dictated by practical limitations, others by the more fundamental restrictions inherent in complex and ill-defined task domains such as road traffic. In Chapter 6 the representation and manipulation of knowledge by the GIDS system are described in detail. The functional architecture of the GIDS system that allow GIDS to behave adaptively in the specified domain is covered in detail in Chapter 7 and the hardware and software structures in which GIDS has been realized are dealt with in Chapter 8. One version of the GIDS prototype is operational as an in-vehicle real-world sysem; a second version operates in a simulated environment. The latter system, which is described in Chapter 9, not only allows more complicated and hazardous behaviours to be studied, but it has the added advantage of providing the possibility of rapid prototyping for future extensions of the GIDS system.

Part III covers the evaluation of the GIDS system's technical and behavioural performance (Chapter 10), its potential impact and acceptance (Chapter 11) and, in conclusion, its importance for the development of RTI technology (Chapter 12). This last chapter closes with some recommendations regarding the further development and application of GIDS.

References

DRIVE (1988). *Call for papers.* Brussels: The European Communities, DRIVE Central Office.

GIDS (1988). *GIDS, Generic Intelligent Driver Support System: A proposal for research.* Haren, The Netherlands: Traffic Research Centre, University of Groningen.

Michon, J.A. (1985). A critical review of driver behavior models: What do we know, what should we do? In L.A. Evans & R.C. Schwing (Eds.), *Human behavior and traffic safety.* (pp. 487-525). New York: Plenum Press.

Michon, J.A. (2 September, 1987). Twenty-five years of road safety research. Opening address to the Second International Road Safety Conference, Groningen, The Netherlands, 1-4 September, 1987. *Staatscourant, no. 168*, 4-6.

Sleeman, D., & Brown, J.S. (Eds.). (1982). *Intelligent tutoring systems.* London: Academic Press.

Chapter 2
Driving: task and environment

Hans Godthelp, Bertold Färber, John Groeger, Guy Labiale

2.0. Chapter outline

This chapter provides a descriptive background of the driving task. Specifically it deals with those aspects of the driving context that are of special relevance to the human driver. This context is structurally determined by the triad driver-vehicle-road (Section 2.1) and functionally by situation-action elements that are revealed by driving task analyses (Section 2.2). In section 2.3 the focus is on road infra-structure, environmental conditions, and traffic management. Here the question is what problems these components of the traffic system are known to create for the human driver and how they might be overcome by various kinds of support. Speci-fically the question is what environment-oriented support will most facilitate driv-ing performance. In Section 2.4 the same question is raised with respect to the ve-hicle: what information is needed for vehicle operation to be as smooth and effective as possible? In the light of the GIDS project it is important to determine what aspects of the driving task are amenable to RTI-technology; after all, smart roads and smart vehicles will be dependent on smart drivers for a very long time to come. This topic is discussed in Section 2.5 and in the final section (Section 2.6) some conclusions are drawn.

2.1 The environment-vehicle-driver system

Together with air and waterway transport the road traffic system serves as an im-portant economic factor in modern society. However, in contrast to, for example, air traffic, the role of human control in road traffic is still relatively large. The im-portance of human control in car driving by contrast with piloting an airplane can be exemplified by the following observations. On the one hand driving under normal circumstances is a comparatively easy task. This conclusion is obvious from the number of lessons a normal person needs in order to acquire the basic driving skills. On the other hand, the variability of driving situations and their correct evaluation is immensely large compared with air traffic situations. This argument is supported by the ubiquitous application of autopilots in all modern airplanes. Autopilots can take over a great many actions because of the predictability of air traffic situations and the clear criteria for their evaluation. Furthermore, it is possible to specify pilot associate systems because of the strict rules applying to flying manoeuvres.

To understand the specific nature of the driving task and the possibilities of RTI systems we must realize that many if not most individual drivers travel from one place to another at self-chosen times, with self-chosen vehicles along self-chosen roads. During their journey they meet hundreds of other traffic participants, which makes driving seem like the task of a bird flying in a flock and finding its way within a myriad of environmental constraints. Nature has equipped birds with elementary sensors which prevent them from colliding despite their high flying speeds. Highly skilled automobile drivers, those who have become one with their vehicle, may reach a similar level of automatic error prevention. Nevertheless, humans may fail for a variety of reasons and consequently road traffic may result in serious accidents and inefficient use of the roadway system. Accidents are seldom attributable to a single cause, but are most often related to a combination of driver, vehicle, and/or environmental factors. Driver failure may result from a lack of skill, which in turn may result from inexperience or from factors such as fatigue or alcohol. The driving environment adds some potential accident causes to human failure. Visibility may be reduced by darkness, rain, or fog. In addition, roadway quality may be limited because of wear and tear, or inappropriate design. Finally, drivers may be confronted with the limitations of vehicle controllability, or with vehicle failure.

Designers of the roadway system have taken many of these factors into account. Visibility problems, for instance, are coped with by modern public lighting and traffic information systems, and roadways are designed in accordance with category-related design rules which allow drivers to behave in an anticipatory way. Nevertheless the damage caused by traffic accidents and pollution is generally judged to be unacceptable, which has led to the conclusion that fundamental innovations are needed to make road traffic suitable for the 21st century. The use of modern information and communication technologies is considered one potential basis for such innovation. A fruitful application of Road Transport Informatics should be based on an evaluation of the accident-related failures of the dynamic triad environment-vehicle-driver to function properly. Analysis of the automobile driver's task in terms of hierarchical task levels and situation-action elements may provide the basis for such an evaluation (Michon, 1985, 1989; Aasman & Michon, 1992).

2.2 Driving task analysis

Task analysis has been used in a wide variety of domains, generally as a forerunner to the development of a performance model of some skill, in order to promote understanding or as a way of supporting performance in a particular domain. A range of task analysis techniques have been employed (Meister, 1985), but virtually all share a tendency to seek to identify the operations involved in performing a task rather than the mental processes which must subserve such operations. There is, therefore, a tendency to attempt to describe observable performance, that is, execution, rather than to identify how such performance is initiated, planned and controlled. This may in part be due to the fact that task analyses tend to be carried out by expert analysts who consider the behaviour of others, or by experts in a particular domain who do not have the intention (or the psychological competence) to carry out a more cognitively oriented analysis. It should be emphasized that this observation in no way detracts from the importance of the careful and authoritative work of, for example, McKnight and Adams (1970 a, b) who are responsible for the most extensive and detailed description of the driving task, originally carried out in order to improve the quality of driver training and driver education in the United States.

McKnight and Adams analysis of what the driver has to do distinguishes 43 separate main tasks which are further broken down into some 1700 sub-tasks. Each of these is further classified into the activities the analyst surmises to be required for such tasks to be performed successfully (Figure 2.1). McKnight and Adams also painstakingly detail available evidence on the likely importance of particular activities indicating, for example, what is known about the contribution which an activity makes to accident causation. As has been pointed out by Groeger (1987), among others, the McKnight and Adams analysis, extensive as it is, is not a theory of learning or instruction. Neither is it a description of how the task is actually performed by the driver. As such we must be cautious about the use to which the results of such an analysis are put, especially where *performance support* is the particular goal to be achieved, as is obviously the case within GIDS. While McKnight and Adams should not be blamed for not considering the processes which underlie performance, as their purpose was not to develop an electronic in-car support system, their task analysis is only of partial use to the GIDS design.

Task 31: Following

		Criticality	
31-1	MAINTAINS ADEQUATE FOLLOWING DISTANCE FROM LEAD VEHICLE		
31-11	Maintains appropriate following distance behind lead vehicle to allow for stopping car in advance of lead vehicle if necessary *	16	X X X X X
31-12	Maintains at least 500 feet separation distance behind emergency vehicles *	0	X X X
31-13	Increases separation distance		
31-131	When following		
31-1311	Oversize vehicles that obsecure forward visibility *	6	X X X X
31-1312	Vehicles which stop frequently (transit and school buses, post office and delivery vans)	11	X X X X
31-1313	Two wheeled vehicles (motorcycles and bicycles) *	6	X X X
31-1314	Vehicles carrying protruding loads *	0	X X X
31-1315	Vehicles driving erratically	9	X X X X
31-132	On wet or icy roads *	13	X X X X X
31-133	Under conditions of poor visibility (see 51, Weather)	12	X X X X X
31-134	In conditions of darkness (see 52, Night Driving)	8	X X X X

Figure 2.1 An example of the McKnight and Adams task

It would perhaps seem obvious to state that training or support should conform to the way in which the task is actually carried out. However, in many cases the detailed characteristics of human information processing in a particular driving task are not known. The implications of this can perhaps be made clear by the following example.

Suppose a driver is approaching a quiet signalized junction, when the traffic lights change from GREEN to AMBER. The purpose of the support system is to aid the driver in making the appropriate STOP or GO decision, failure to do so being a major contributing factor in accidents at intersections. Accounts of how drivers perform such a task appeal to some notion which involves the driver assessing when the junction will be reached (see Allsop, Brown, Groeger, & Robertson,

1990). It is at least conceivable that such assessments may be made either by estimating the distance to be travelled and the approach speed, or by somehow detecting the rate of expansion of the image of the junction on the retina and estimating the time which would elapse before it is reached. The latter explanation,
based on Gibson's (1968) ecological optics, has enjoyed some currency but is chal-
lenged by the recent work reported by Cavallo and Laurent (1988) and by Groeger
and Cavallo (1991), who show evidence for the distance/speed explanation. Whatever the means by which drivers assess when the junction will be reached, it should
be clear that providing the driver with extra information about actual approach
speed or actual distance to be travelled is unlikely to be beneficial if information
directly available from the changing optic array is used for speed or distance perception. Similarly extra information about the time it will take to reach the intersection is unlikely to be helpful if the driver's real difficulty lies in accurately assessing the distance to be travelled. Also, it is conceivable that such support might
prove damaging if, for example, the driver's focusing on an in-car screen disrupts
the maintenance of some visual or spatial image of the external scene, as may auditory messages such as "two seconds" or "100 metres" if the driver cannot gauge
exactly how such physical parameters relate to the external scene. Simple STOP or
GO advice would only be appropriate if the reaction time and speed preferences of
the driver could be taken into account.

This serves also to illustrate some other shortcomings of the traditional task description approach. It is known, for example, that factors such as age, fitness, goals
and time pressure all distort reaction times. An indication of the time remaining before the junction will be reached, if it is to prove helpful, must therefore take into
account variations within and between individual drivers. Generally, task analysis
methods fail to explicate how such individual differences, or how a choice of alternative ways of meeting the demands of the task, can be included. Such analyses
also fail to incorporate the fact that the way in which the task is performed by inexperienced and experienced drivers may fundamentally differ (see Cavallo &
Laurent, 1988). Obviously, only an account of the processing underlying the activity to be performed can meet such needs.

Given the difficulties inherent in interpreting and applying the traditional formal
task analysis of the driving task, it would be easy to conclude that this approach is
of little use in the development of a support system such as GIDS, the hallmark of
which is adaptability to changing circumstances and driver needs. But this is not
the case. The McKnight and Adams account of the tasks drivers must perform remains the most detailed and comprehensive description of this task domain and it
has been used extensively, as will be seen in later sections of this book, in the initial task description and design of some components of the GIDS prototype. As
will also be clear from subsequent sections, the most serious shortcoming of the
McKnight and Adams work, namely the failure to provide adequate behavioural
corroboration of the task descriptions formulated, has also been addressed, and as a

result proposals for supporting drivers are based on empirical data collected by various teams within the consortium, rather than on arbitrary accounts of the driving task.

Actually the approach of GIDS has been to provide a basic reference model which quantitatively describes the optimum driver action and vehicle motion characteristics for a particular manoeuvre. This model takes account of driver, vehicle, and traffic characteristics and, as such serves as an adaptive, normative description of behaviour. Thus, aspects like driver age and experience, vehicle weight and dimensions, and traffic density become the natural components of the driver support system, which is meant to help the driver's navigation, manoeuvring and vehicle control performance in a variety of circumstances.

2.3 Infrastructure

Many problems that traffic participants encounter on the road are caused by the fact that the existing roadway infrastructure and traffic control system have been suboptimally designed. Actually the roadway system in most Western countries is based on a mixture of historical and modern design principles. Modern highways and new towns are connected with a road network which, in many cases, has its origins in the nineteenth century. Consequently drivers, cyclists, and pedestrians are confronted with a variety of not quite compatible traffic circumstances which makes their task quite complex.

Currently roads are designed in accordance with a series of design rules that belong to a certain road categorization scheme. Specific behavioural rules, for instance about speed limits, are associated with each road category. Drivers' expectations about the chance of meeting other traffic are also governed by the design principles for a special road category. Theoretically this design philosophy ought to result in almost perfect driving behaviour. However, in practice several problems may occur. In many cases drivers can hardly distinguish between different road categories. On the one hand, this is caused by ambiguity in the use of specific category-related design features. On the other hand, local circumstances often force the road engineer to use substandard solutions: for instance, relatively short sight distances. Furthermore, the question arises whether so-called optimal technical solutions for, say, intersection design are really optimal from the driver's point of view.

That all is not well seems to be evident from the fact that various information systems have been developed to provide additional information to the road user. Road signs warn of oncoming hazards and give information about local traffic rules in terms of speed limits, right of way, etc. In many cases these signs are necessary to compensate for suboptimal design charactics. The modern use of signs and signals may also form an integrated part of a specific road category. Route signs that are placed well in advance will help the driver to behave in an anticipatory way. When approaching an intersection, vehicle-actuated traffic signals may serve as an intelligent tool which controls the right-of-way rule in a traffic-dependent manner. Road delineation systems not only guide the driver along the road, but may also give information about local overtaking possibilities. Driving performance on a suboptimally designed road may reach a critical level in darkness and/or bad weather. Even for an experienced driver, vehicle control may become a difficult task in conditions of heavy wind or on a slippery road. In many cases such circumstances occur in combination with bad visibility, because of rain, fog, or snow. The relatively high number of bad-weather accidents and the seriousness of these accidents indicates that drivers are not well prepared for such conditions. Apart from general warning signs which indicate the potential occurrence of rain or fog, the road infrastructure usually provides little help on this point. Nevertheless, in their integrated form modern road designs and traffic information systems may give the road infrastructure an almost perfect appearance. Particularly in such a case, it is important to realize that many parts of this infrastructure are not suited to meet the demands placed on them by the (imperfect) road user. Traffic participants do and will continue to make errors. A future, more safe and efficient infrastructure should therefore be based on roadway design rules and supportive driver information systems that can help to avoid or compensate for these errors.

GIDS is designed such that roadside and in-vehicle information sources can be combined in a way which takes account of driver characteristics. Traffic-dependent route information via variable message signs may well be integrated with in-vehicle *navigation* support. Roadside advisory speed and headway warnings should be designed in combination with in-car intelligent cruise-control and collision-avoidance systems. These systems may be applied together with green-phased traffic signals and speed limitation systems for safe and efficient traffic control and driver *manoeuvring*. Lane crossing, because of narrow lanes and/or driver inattention, may be supervised and corrected by an integrated roadside/in-vehicle warning system for proper *vehicle control*. GIDS should manage the information stream related to these navigation, manoeuvring, and control support functions in a user-friendly way.

2.4 Vehicle

Vehicle control constitutes the basic task element in driving. Following the roadway curvature and path control between other traffic participants and obstacles require correct steering and speed control actions. Vehicle stability and controllability is largely speed dependent. In other words, speed and steering control should be carried out in a highly correlated manner. Drivers may have problems in adapting speed in an anticipatory way in order to prevent traffic conflicts. Active warnings alongside the road or in-vehicle can be used to give immediate feedback about the discrepancy between actual and safe speed.

In future vehicles, steering and speed control will be combined with the supervision and handling of a complex set of in-vehicle instrumentation, that is, route guidance, radio, telephone, etc. The introduction and wider use of such systems may result in dramatic consequences if their design is not based on knowledge about the factors influencing driver information processing capacity and workload. Regarding information presentation and control, the questions should be considered of what mode (visual, auditory, tactile) and at which moments information should be presented, and how natural and artificial intelligence should interact.

Information presentation

To drive safely, drivers have to pay close attention to the road scene, in order to keep their vehicle on the correct trajectory, to avoid obstacles, to read road signs, to respect traffic rules, etc. Consulting in-vehicle displays may conflict with these driving requirements, particularly if the design of these devices has not been ergonomically defined. Deficient design and organization of such displays may cause various sorts of problems:

- difficulties in searching for or reading information, thus increasing duration and number of visual glances away from road scene, and increasing attentional effort;
- difficulties in understanding symbols and information, thus causing misunderstanding and confusion;
- difficulty in taking into account simultaneously a great amount of in-car data, thus increasing the mental workload of the driver. As a consequence, bad designs of in-car displays may disrupt the driving task, causing near-accidents (such as lane deviation, violation of traffic rules) or actual accidents;
- to circumvent these problems, an intelligent structure has emerged, represented by the GIDS system. The purpose of this intelligent interface is to integrate new information and communication functions into cars in order to manage the ergonomics of the various messages presented (or the type of control required) their presentation and the personal characteristics of each driver. The main objectives of an intelligent interface are adaptation to a specific context, understanding what drivers require, reasoning from pre-assimilated information and the acquisition of new information. The aim is to facilitate driver understanding, by presenting information in an appropriate form, taking account of the specificity of each user, each situation and their needs.

Basic issues on dialogue control and information management

It should be emphasized that whilst conventional dashboard ergonomics (e.g., Labiale, 1990) takes into account a *surface ergonomics* of the vehicle displays, the intelligent interface of GIDS uses not only this surface interface, but also requires new ergonomic developments, which we might call *cognitive ergonomics*. Cognitive ergonomics requires definitions of new criteria (Labiale, 1991) concerning information management by an intelligent co-driver, such as GIDS. These criteria concern:

- absence of competition with other sources of visual data on the dashboard and for other media such as audio messages and manual controls;
- relevance of the content and timing of the data supplied in terms of the road situation and drivers' expectations;
- adaptation to the personal characteristics of drivers (e.g., experience, age, handicap, driving style) and to the manner in which drivers request information and dialogue;
- respect of personal preferences for independence;
- reliability of data supplied, enabling the creation of driver confidence.

This set of requirements has led the GIDS consortium to consider a broad spectrum of interface components as potential information systems, including visual displays, voice generation, and active control. Proper implementation of non-visual information systems, distribution of messages over time and adaptation to the skills of individual drivers may together keep driver workload low and facilitate information processing. Defining information management and dialogue structure along these lines may, furthermore, ensure that drivers will accept such support functions and that safety and comfort are improved.

Controls

Most aspects mentioned for information presentation are also relevant for control design. However, two main differences should be noted, one positive and one negative. Whereas the presentation of information on displays is mainly externally triggered (by the car or the environment), controls are activated in a driver-paced fashion. This means that drivers are better able to decide at what time they wish to focus their attention on the control inside the car. Unfortunately, however, the mean search and handling times exceed those for attending to displays.

A study by Burger et al. (1977) concludes that at least 7.5 per cent of all accidents result from a mismatch in the driver/vehicle system. So, the question arises, what the characteristics are of these mismatches. To understand the problems of poorly designed controls one must consider the fact that controls have an actuating and a feedback function. With respect to the actuating function, controls must be positioned within the optimal reaching area of the driver. But this area is quite small, because of the length of the driver's arms and legs and restricted space in the car (cf. ISO 3436.3 and ISO 3958). To optimize the feedback part of controls, they should be visible and self-explanatory. Self-explanatory means, for example, a toggle switch in the up-position is set to 'on', in the down-position to 'off' (at least, that is what continental European designers consider 'self-explanatory'!).

The attribute 'self-explanatory' implies that every driver will spontaneously operate a specific control correctly without training. Self-explanatory properties of controls may further be improved by providing active feedback via a particular manipulator (Godthelp, 1990). Force feedback via the pedals and steering wheel have been adopted, in principle, as components of the GIDS human-machine interface. These active controls can provide information about occasional speed or lateral path errors. Intelligent controls will thus be a part of an adaptive information system which improves vehicle control and reduces related workload levels.

2.5 Road transport informatics

The question what benefits can be expected from Road Transport Informatics (RTI) systems leads to totally different answers and predictions by different experts. Ad hoc evaluations of safety benefits from new vehicle technologies reveal a wide range of results for similar systems from different experts and also great variety between the functions (PRO-GEN Safety Group, 1989). The main dispute between the various experts as to whether RTI systems can reduce accident rates by 5, 10 or more per cent is inappropriate. A more consistent evaluation pattern requires several distinctions.

The first important aspect to bear in mind is that RTI systems can, though not necessarily do, help the driver to perform better by reducing errors and accidents. This means that one must take into account that drivers are not only information-driven, but also motivation-driven. RTI systems are especially helpful where information deficits lead to wrong or inappropriate behaviour. They are almost useless, however, when motivational factors dominate in dangerous or incorrect driving behaviour. Acceptable and accepted RTI systems must be compatible with the intentions of the driver. Unreliability and low acceptance will frustrate wide distribution, which is itself a precondition for the success of such systems. It should be noted at this point that most estimates of the success of RTI systems presuppose a high equipment penetration. Also, reliability problems are often ignored. Theoretically, reliability is treated as a precondition for installing a new RTI system. However, everyday experience with new cars shows that the reliability of new technical systems is far below the desirable level. Anti-collision or distance-keeping systems are prominent examples. Currently these systems guarantee high reliability only at the price of false alarms, which themselves cause acceptance and safety problems. A system that warns unnecessarily often and too early will be switched off by the user and therefore be useless!

To understand and evaluate possible benefits of RTI systems in general and of the GIDS system in particular, a categorization of the driving task is necessary. Following the categorization made in Section 2.2, the driving task can be structured at three levels. These levels are navigation, manoeuvring, and control. At the navigation level drivers choose their destination and desired route. Their choice may depend on specific goals or traffic conditions. They may, for example, choose different routes if they are on a holiday or a business trip, if they desire to admire the scenery or to travel the fastest way. It is obvious that such choices will influence their information needs. Because human perception and information processing are conceptually- rather than data-driven processes (cf. Norman & Bobrow, 1975), the decision to take a specific route changes the information flow, the load and the benefit of an RTI system. During their trip drivers will meet other cars, pedestrians, cyclists, etc., and they must relate their intentions and actions to those of other road users. This level of action is termed manoeuvring. Typical actions are distance keeping, lane changing, overtaking, passing other vehicles or crossing an intersection. The lowest level of car driving behaviour, handling and control, covers actions like steering, acceleration, and deceleration. At this microlevel, a driver's actions are influenced by actions at the manoeuvring and navigation level, and also by the reactions of the car and the sensations and perceptions of the driver. Single accidents, like veering off the road or driving too fast on a bend, are typical examples of faulty control behaviour. It should be emphasized that at this level the coupling between perception and action is very important in guaranteeing immediate and correct reactions by the driver. In other words, the value of RTI functions, such as lane keeping support, is intrinsically related to their ergonomic design (see Sections 4.6 and 5.1). While the direct impact of wrong reactions at the handling or the manoeuvring level is widely accepted, the influence of navigational problems on traffic safety and accident rates is often ignored. However, as Färber et al. (1986), and Engels and Dellen (1989) have shown, drivers who are unfamiliar with the locale and/or disorientated cause more accidents or traffic conflicts than drivers who know where they are and where they need to go.

A further and equally important aspect is inter-individual variability. A review of the literature on work psychology reveals that it is shortsighted to 'optimize' work environments with regard to a 'standard' person. Extrapolating this insight to RTI systems, we must consider individual differences as an important factor. A function that is helpful to one driver (e.g., distance keeping) can annoy and, as a consequence, disturb others. Support systems must therefore be adaptive.

The final important aspect of evaluations of the benefit of RTI systems is how to assess the overall safety of a combination of functions. As PRO-GEN (1989) pointed out, "to estimate the potential impact on safety of a group of these functions, it is not possible simply to add up the separate potentials." Or, as Gestalt psychology stipulated, the whole is more than the sum of its parts. In order to make intelligent use of and obtain maximum benefits from RTI systems we must take into account mutual interactions. Interactions can be positive in the sense of synergetic effects, but also negative or antagonistic. So, what do we know at present about the potential impact of RTI systems on traffic safety? We know that it depends on several variables. Only a system which takes all these variables into account will use RTI systems in a productive manner.

2.6 Conclusions

In summary it can be concluded that the application of modern communication technologies to road traffic requires a careful consideration of drivers' information needs. The driving task is a complex combination of subtasks, each of which may form a basis for the commission of fatal errors. Driver support systems should therefore be designed such that the major subtasks in driving, that is, navigation, manoeuvring, and vehicle control are dealt with in an integrated manner. A detailed description of drivers' potential errors and their underlying causes should form the basis for a normative driver model which serves as a reference for the appropriate warning strategies. Warnings should be self-explanatory and presented in accordance with rules derived from modern cognitive ergonomics. Proper implementation of non-visual information systems, distribution of messages over time and adaptation of support to the skills and expectations of individual drivers should keep the workload low and facilitate information processing.

Ultimate safety and efficiency will depend on drivers' motives and intentions. RTI systems will be effective only if they help the driver in a way which is recognized and accepted by individuals and by society.

References

Aasman, J., & Michon, J.A. (1992). Multitasking in driving. In J.A. Michon, & A. Akyürek (Eds.), *Soar: A cognitive architecture in perspective* (pp. 169-198). Dordrecht, The Netherlands: Kluwer.

Allsop, R. E., Brown, I. D., Groeger, J. A., & Robertson, S. A. (1990). *Modelling driver behaviour at actual and simulated traffic signals.* Contractor's report CON/9834/35. TRRL, Report CR 264. Crowthorne, United Kingdom: Transport and Road Research Laboratory.

Burger, W., Färber, B., Queen, J., & Slack, G. (1977). *Accident and near accident causation: The contribution of automobile design characteristics.* Report NHTSA. Springfield, VA: National Technical Information Service.

Cavallo, V., & Laurent, M. (1988). Visual information and skill level in time-to-collision estimation. *Perception, 17*, 623-632.

Engels, K., & Dellen, R.G. (1989). Der Einfluß von Suchfahrten auf das Unfall-verursachungrisiko [The effect of searching while driving on accident risk]. *Zeitschrift für Verkehrssicherheit, 35, 3*, 93-100.

Färber, B., Färber, Br.A., & Popp, M. (1986). Are oriented drivers better drivers? *Proceedings 5th International Congress ATEC*, 'The lack of road safety', Vol. 7 (pp. 1-9). Paris: ATEC.

Gibson, J.J. (1968). What gives rise to the perception of motion? *Psychological Review, 75*, 335-346.

Godthelp, J. (1990). *The use of an active gas-pedal as an element of an intelligent driver support system; literature review and explorative study.* Report 1990 B-17, Soesterberg, The Netherlands: TNO Institute for Perception.

Groeger, J.A. (1987). Underlying structures: Driver models and model drivers. In J.A. Rothengatter & R.A. de Bruin (Eds), *Road User Behaviour: Theory and Research* (pp. 518-526). Assen, The Netherlands: Van Gorcum.

Groeger, J.A., & Cavallo, V. (1991). Judgements of time-to-collision and time-to-coincidence. In A.G. Gale, I.D. Brown, C.M. Haslegrave, I. Moorhead & S. Taylor (Eds), *Vision in Vehicles III* (pp. 27-34). Amsterdam: North-Holland.

ISO 3409 Edition (1975). TC 22. *Passenger cars - Lateral spacing of foot controls.* Genève: ISO Central Secretariat.

ISO 3436.3 Edition (1980). TC 22. *Road vehicles - Controls, indicators and tell-tales - vocabulary.* Genève: ISO Central Secretariat.

ISO 3958 Edition (1977). TC 22. *Road vehicles - Passenger cars - Driver hand control reach.* Genève: ISO Central Secretariat.

Labiale, G. (1990). *Psycho-ergonomie de l'interface conducteur/automobile* [Psycho-ergonomics of the driver/car interface]. October 1990 (pp. 22-26). Revue Génerale de l'Electricité.

Labiale, G. (1991). *L'interface intelligente dans la voiture.* [The intelligent interface in the car]. Scientific Communication at the colloque on "L'intelligence artificielle et les sciences humaines". Lyon: University of Lyon II.

McKnight, A.J., & Adams, B.B. (1970a). *Driver education task analysis.* Volume I: Task descriptions. Final report, contract no. FH 11-7336. Alexandria, VA: Human Resources Research Organization.

McKnight, A.J., & Adams, B.B. (1970b). *Driver education task analysis.* Volume II: Task analysis methods. Final report, contract no. FH 11-7336. Alexandria, VA: Human Resources Research Organization.

Meister, D. (1985). *Behavioral analysis and measurement methods.* New York: John Wiley.

Michon, J.A. (1985). A critical review of driver behavior models: What do we know, what should we do? In L. Evans & R.C. Schwing (Eds), *Human behavior and traffic safety* (pp. 485-520). New York: Plenum Press.

Michon, J.A. (1989). Explanatory pitfalls and rule-based driver models. *Accident Analysis & Prevention, Vol. 21, 4,* 341-353.

Norman, D.A., & Bobrow, D.G. (1975). On data-limited and resource-limited processing. *Cognitive Psychology, 7,* 44-64, 1975.

PRO-GEN safety group (1989). Estimation of the potential safety effects of different possible PROMETHEUS functions. Stuttgart: PROMETHEUS Office.

Chapter 3
The driver

Talib Rothengatter, Håkan Alm, Marja J. Kuiken, John A. Michon, Willem B. Verwey

3.0 Chapter outline

This chapter deals with the human organism as a component in the traffic system. With other components relatively invariant over time, most of the burden of adaptation to the fast changing conditions in road traffic falls on the flexible human being. This leads us to a brief review of driver characteristics in Section 3.1, with an eye on those aspects of these characteristics that are to be formalized such that they can be implemented in a driver support system such as GIDS. This question returns in the subsequent two sections. First in Section 3.2 we look at the driver's task structure, in an attempt at converging on a task description that is computationally suitable. In Section 3.3 the relevant driver information needs will be specified, and in Section 3.4 the outcome of this analysis will be brought to bear upon the important concept of driver workload. How is it measured, and how can it be quantified in such a way that it is possible to use workload as an active variable in the GIDS system? The conclusions reached on the basis of these analyses will be summarized in Section 3.5 and serve as the starting point for the discussion of driver support in Chapter 4.

3.1 The analysis of driver characteristics

The GIDS system will provide support in driver navigation, manoeuvring and control tasks. While using GIDS for doing each of these tasks, the physical, perceptual and cognitive characteristics of the driver need consideration. In terms of physical capacity, the design and layout of controls must be such that they can be easily reached and operated by all drivers. Here anthropometric data on various population groups must be considered and, specifically, elderly drivers, as these appear to be particularly prone to performance decrements due to poor ergonomic design of sensori-motor tasks (e.g., Welford, 1982). In terms of perceptual capacities, consideration must be given to such matters as legibility of visual material and clearness of auditory presentations under various conditions, for instance where glare or low light is present or where noise levels outside or inside the vehicle are high, for drivers of different ages and sensory capacities.

In terms of cognitive capacities, consideration must be given to information processing rates in different traffic situations for drivers of various levels of experience and of different ages, short-term memory capacities, effects of expectation on response (population stereotypes), and decision making in the face of the inevitable false alarms and missed signals in a GIDS equipped vehicle. In terms of cognitive task representations and thought processes, finally, consideration must be given to the various strategies (or schemata) that people employ to perform various aspects of the driving task. In navigation, for instance, the degree to which a particular driver is familiar with the local circumstances influences route finding in specific ways that seem to lend themselves to particular driver support techniques. At a high level of familiarity a mental map is available that can be read off directly. If no consistent mental map is available, but the driver is familiar with the local topography, a form of cue-based navigation takes precedence. This is superseded by either a dead-reckoning or map-reading strategy if little or no local topography is represented (e.g., Rothengatter, 1989).

It has been argued that many drivers have an unrealistic view of their own driving ability. Finn and Bragg (1986), Matthews and Moran (1986), and Svenson (1981) argued that the majority of drivers believe themselves to be more skilful than the average, although Groeger and Brown (1989) conclude from their own study of this phenomenon that the reported effects may be largely artifactual. Näätänen and Summala (1974, 1976) stress the importance of the driver's motivation to drive safely, arguing that drivers often have different motives such as sensation seeking and competition, which will motivate them to engage in risky behaviour. Although the ways drivers 'perceive', negate, take and compensate risk is important in explaining accident involvement, the GIDS system does not need to deal directly with the concept of risk. Risky behaviour is detected by the support system in terms of overt behaviour and is therefore dealt with directly without referring to the conceptually vague notion of risk.

Finally, an important factor influencing driver behaviour is the driver's momentary state. Thus, trip purpose (e.g., private versus business trips) has a significant effect on intentional speed choice. Drivers involved in stressful personal situations have been shown to suffer increased accident involvement. Aggression resulting from frustrating traffic conditions is the subject of much anecdotal information, but there is no firm empirical evidence that 'aggressive' driver behaviour is a frequent occurrence (Rothengatter, 1991). Alcohol and drug usage strongly affect driver performance and accident involvement (for review see Smiley & Brookhuis, 1987). Driver vigilance is affected by time-on-task, sleep deprivation and fatigue (O'Hanlon, 1981), which can be assessed on the basis of vehicle parameters (Brookhuis, 1990). Although the GIDS system cannot consider such momentary fluctuations in the driver's state directly, it can take account of changes in driver state insofar as these are observable through changes in the overt behaviour of the driver. How these observable changes are taken into account by the actual driver model incorporated in the GIDS system is described in more detail in Part II of this volume.

Stable differences in driver characteristics are documented more extensively. Demographic variables such as age, sex, occupation or income are used to classify certain driver populations for predicting the likelihood of problem behaviour such as drinking-driving. In the study of speed choice characteristics such as car purchase preferences, journey purpose, annual mileage, and occupation have been used to predict habitual speed choice. Experience appears to be related to reported differences in driving style and the occurrence of driver errors (Groeger et al., 1990). Maycock, Lockwood, and Lester (1991) identified age, number of years since obtaining driving licence and annual mileage as the most important determinants of accident involvement. Evans (1991) discusses several studies reporting a higher accident involvement of males relative to females, concluding that this reflects a greater activity level and/or greater propensity to take risk by males compared with females.

For a driver support system, the relevant driver characteristics are those that not only reflect a difference in accident involvement but also a distinct difference in traffic behaviour. This has been well documented in the case of age and experience and, to a lesser extent, for sex. For example, observed driving speed and tendency to follow closely are significantly influenced by age and experience. For this reason, the GIDS system focuses initially on these two variables, with the option to include other relevant variables whenever sufficient empirical information becomes available to do so in a meaningful way.

Age

Age and experience effects are naturally confounded variables when young drivers begin to participate in traffic independently. The few who start to drive at a later stage are in so many ways atypical that they are unsuitable for comparative studies. This is not so for later stages in life. Here specific age effects can be identified irrespective of driving experience. Elderly drivers (above 60 years of age) appear not only to be disproportionally involved in road traffic accidents, but are also more often legally at fault (Engels & Dellen, 1983; Fontaine, 1988). In an empirical study with different age groups Brendemühl, Schmidt, and Schenk (1988) found significant increases in errors in perception of priority regulations, perception of traffic signs and lane keeping with increasing age. Van Wolffelaar (1988) concludes that specific functional deficits can be identified as contributing factors. These include (i) perceptual determinants, in particular selective attention processes; (ii) decisional determinants, in particular divided attention processes, and (iii) motor determinants. Experimental studies investigating the effects of introducing RTI as additional driving tasks demonstrated dramatic effects on both sensorimotor processes (lane keeping) and perceptual processes (detection of peripheral stimuli) which were to a large extent age-dependent. Particularly detrimental effects were found in paced presentation of information, short stimulus durations, and manual response modalities (Van Wolffelaar, Brouwer, & Rothengatter, 1990). Similar effects were found in a study investigating the use of cellular telephones while driving (Nilsson & Alm, 1991). Van Wolffelaar et al. concluded that, as the effects of interactions in RTI-related tasks depend to a large extent on age-specific capacities of the driver, it will be necessary to evaluate all RTI implementations that require some form of interaction during the execution of the driving task separately for drivers of different age groups. Moreover, since some modes of presentation and response seem to be more favourable for young drivers and others more favourable to elderly drivers, it is recommended that RTI implementations are adapted to the specific age of the driver. Even though elaborate classification schemes have been proposed in the framework of RTI (e.g., Warnes, Frazer, & Rothengatter, 1991), here it suffices to distinguish roughly three age groups, young (18-24 years of age), adult (25-65 years of age) and elderly (65+), as further specification would be meaningless because of the large within-group differences amongst the young – where age and experience are confounded – and the elderly – where cognitive ageing deviates from chronological ageing.

Experience

Drivers with different levels of experience are not all equally likely to be involved in an accident. The inexperienced young driver, for example, is much more likely to become involved than his experienced counterpart. Findings of a study by Pelz and Schuman (1971) showed that accident rates rise to a peak some two or three years after young drivers obtain their licence. Findings from various studies suggest differential involvement in accident types. Van Kampen (1988) analyzed the accident involvement of inexperienced young drivers and showed that these drivers, compared with more experienced drivers, were more often involved in accidents when taking a bend and less often when negotiating an intersection or junction. Similar findings were reported by Haas and Reker (in Groeger, 1991); experienced drivers were more often involved in accidents caused by violating the regulations or by lack of attention, whilst inexperienced drivers' accidents arose from excessive speed, lack of proficiency in traffic and faulty overtaking.

    Successful skill acquisition and skilled performance will depend to a greater extent on knowledge about the success of previous activities; 'feedback' has a crucial

role to play in error- and accident-free behaviour. There is no reason to believe that
this analysis is not equally true of driving. Unfortunately, the feedback available to
drivers when they have ceased formal training is very limited. Supplementing the
feedback on performance available to drivers would, in principle, enhance driving
skill (though not equally) across all age/experience groups. However, the greatest
benefit from feedback is likely to accrue where the information given matches the
individual needs of the performer. From this point of view 'adaptability' of support
to individual needs is crucial.

The level of experience is defined here by taking into account the number of kilometres driven, the number of years the subject had a driving licence, and the
number of kilometres driven during the last 12 months. This leads to the categorization presented in Table 3.1.

Table 3.1 - Definition of driving experience

Inexperienced novice	licence < 1 yr driven < 1000 km
Experienced novice	licence < 1 yr driven > 1000 km
Inexperienced driver	licence 1-5 yrs driven < 100 000 km/last 5 yrs or: licence > 5 yrs driven < 10 000 km/last year
Experienced driver	licence < 5 yrs driven > 100 000 km/last 5 yrs or: licence > 5 yrs driven < 100 000 km/last 5 yrs and > 10 000 km/last year
Very experienced driver	licence > 5 yrs driven > 100 000 km/last 5 yrs

Although the relative changes in accident involvement with increasing driving
experience are well documented, it is less than clear why these changes occur. To
some extent they can be attributed to changes in exposure. As mentioned above, inexperienced drivers can be more or less equated with young drivers, who tend to
have different trip purposes than older drivers. For example, they drive more at
night, and night-time driving in general is more risky. They also have, in general,
smaller and older cars, which increases the injury likelihood in accident involve-

ment (see Evans, 1991). Younger drivers also are more likely to indulge in drink-ing-driving and are less likely to wear a seatbelt, which again increases injury like-lihood when involved in an accident. Hence, the increased accident involvement found amongst inexperienced drivers may to a considerable degree be attributable to different exposure to injury risk characteristics.

Changes in performance have been reviewed by Groeger (1991) who reports two studies, one carried out by Quenault and Parker (1973), comparing novice drivers, drivers with one year's experience and 'normal' drivers, reporting consid-erable differences between these groups, and another by Duncan, Williams, and Brown (1991) which fails to find consistent differences between drivers with dif-fering levels of experience. Unfortunately, in both studies the extent of experience and training was not controlled, which led Groeger to conclude that speculation as to whether an adequate level of performance has deteriorated or has never been achieved, is premature. Kuiken (in Kuiken & Groeger, 1991) carried out an ex-perimental study comparing amongst other factors, high and low mileage, novice versus experienced and inexperienced drivers. She found that these groups do dif-fer on subjective effort scales and self-assessment of driving quality, but no dif-ferences were found for objective behavioural measures such as negotiating bends, approach to intersections and lateral control of the vehicle. In a second study, Mil-tenburg and Kuiken (1991) compared visual search strategies in groups with differ-ent driving experience. In this study the hypothesis that experienced drivers have a shorter duration of the first fixation on relevant cues was not supported. Neither could it be concluded that experienced drivers fixate more quickly on informative (relevant) cues than less experienced drivers. Finally, no differences were found in the distance of eye fixations from the driver.

It is obvious from these studies that the classification of 'experience' is problem-atic. Time since acquisition of driving licence may be an accurate predictor of acci-dent involvement probability on an aggregate level. It appears less useful, however, for determining individual performance. Much more research is required in this area, and it is questionable whether experimental on-the-road studies are the best approach. On the one hand, if optimal performance indices are required it might be better to test subjects in a controlled environment, presenting extreme situations. On the other hand, if differences in 'driving style' are to be determined it may be better to use more non-intrusive observational studies. Such a distinction between what individual drivers can and cannot do and what they are likely to do given a certain set of traffic conditions is extremely relevant for the design of the GIDS system. The performance of a driver indicates when and in which circumstances they are likely to commit errors and hence would be in need of support or, consid-ered in the longer term, tuition. Here the conclusion is warranted that both the young, inexperienced and the older drivers will need support more often and at an earlier stage than experienced drivers. The exact situations that warrant support are different for these two groups and so is the type of support required. These differences are relatively well documented. A more difficult issue is the role of the GIDS system in relation to habitual driving behaviour. For example, the finding that younger, less experienced drivers choose to drive with reduced headway may be an argument to increase the threshold for these drivers for an anti-collision device, which would be in line with the adaptivity characteristic of the GIDS system, reasoning that in such cases a reduced headway cannot be considered an error because it represents intentional behaviour on the part of the driver. On the other hand, from the point of view of an optimally safe traffic system, such behaviour, albeit intentional and habitual, has to be considered as erroneous because it decreases the safety performance of the system as such. A similar argument holds for other habitual behaviour such as speed choice. The argument becomes even more cogent if mode and route choice are considered. Should the support system switch on every time the driver starts the car and present the message that it would be safer to take the train? It might be conceivable to advise the driver against travelling certain notorious routes on certain dates, but is it equally conceivable to warn the driver against driving in certain EC countries at all, simply on the basis of relative accident statistics? Warning the driver against his habitual, intentional behaviour is not considered to be an explicit support function, even when such behaviour may constitute an increased accident risk and/or deviate from what is legally required. For this reason, the habitual behaviour of the individual driver is taken as a de facto norm for that driver in the GIDS system, even if this implies that the driver may encounter irrecoverable error situations. Other systems, such as AUTOPOLIS, are designed with the explicit function to ensure that the driver is complying with the legal regulations (e.g., Rothengatter & Harper, 1991). AUTOPOLIS may be considered as a sub-system to the GIDS system, but does not, at present, constitute part of the latter.

3.2 Analyzing the driver's task

Driving, especially in the presence of other vehicles, requires the performance of a non-trivial set of subtasks of the driving task. These tasks can be subjected to a task requirement analysis, specifying in detail the tasks to be performed and the subtasks required for adequate performance. For driving, such an analysis has been carried out by McKnight and Adams (1970) for the purpose of developing a driver education curriculum. However, the main shortcoming of such an analysis is that it ignores the fact that certain tasks or subtasks have to be carried out continuously whereas other tasks have a discrete nature and that, hence, at times simultaneous or parallel task execution is necessary. Michon (1985), for this reason, proposed to follow a production system approach to driver task modelling, which would offer the flexibility required for hierarchically-structured performance characteristics for driver behaviour. For the purpose of taking into account individual characteristics, it is relevant to distinguish main task levels in terms of navigation, manoeuvring and vehicle control and, in particular, to distinguish between knowledge-based, rule-based, and skill-based behaviour (see Michon, Smiley, & Aasman, 1990). Whereas the experienced driver may execute a certain task as skill-based behaviour and is therefore prone to a form of error that is due to monitoring failure, the inexperienced driver may be executing the same task on a knowledge-based or a rule-based level and may be prone to error due to an inappropriate or inadequate problem-solving strategy. Since such errors would need different correction strategies, GIDS needs to infer at which level the driver is performing the task and, hence, needs to knows his or her level of experience. At all levels many errors are due to incorrect expectancies in the sense that the incorrect production or problem-solving strategy is applied to the situation at hand, or that the production or strategy is not adapted to the changing situation. Here GIDS can fulfil an essential support function in alerting the driver to crucial changes in the momentary traffic situation. In order to be able to do so appropriately, the GIDS system needs to have a formalized description of the relevant tasks available at a level of explicitness that is sufficient for a computational model, which in practice means that the task descriptions are formulated in terms of IF...THEN...ELSE statements incorporating all the knowledge and rules of operation that are required to perform the task specified correctly (see Chapter 6). This also means that these requirements are in reality only achievable for certain subsets of the driving task and for this reason the 'Small World' paradigm was conceived. This Small World is discussed at length in Chapter 9.

3.3 Driver information needs

To drive safely drivers must be able to perform a number of basic information processing tasks. First of all, they must be able to perceive relevant elements in the traffic environment. A basic level of perceptual ability is needed, and also a vehicle and a traffic environment that make it possible to perceive the most important environmental features. Due to the enormous range and complexity of traffic situations, drivers must also be able to categorize those features that are functionally similar (with respect to their task demands). The advantage of this categorization is that it will speed up the processing of information. Drivers must also have basic knowledge about different traffic situations, frequently referred to as scripts, plans, or scenarios (Schank & Abelson, 1977). This makes it possible to predict what to look for in different categories of traffic situations, where to look, and when to look for it. From these observations drivers must be able to predict the future behaviour of other road users. Based upon these predictions they must also be able to decide on suitable responses to the predicted behaviour of other road users. Furthermore, drivers must be able to implement these responses as actual behaviour, supervise the responses of other road users and, if needed (if there is a large enough deviation be-tween predicted and actual behaviour of other road users), adjust their own behav-iour. Drivers must also have a realistic view of their driving ability. Finally, they must know what is and what is not safe driving, and have the capability and moti-vation to drive in a safe way.

Research has revealed correlations between many of these information pro-cessing tasks and accident involvement. There is, for example, a weak correlation with certain perceptual abilities (e.g., Hills, 1980; Treat, 1980). A difference in vis-ual search patterns between experienced and novice drivers has also been reported (Brown, 1982). Brown suggests that this might partly explain the higher accident involvement of young drivers. Rumar (1990) claims that late detection is probably a cause of many accidents. There is also a correlation between selective attention and accident involvement (Kahneman, Ben-Ishai, & Lotan, 1973; Avolio, Kroeck, & Panek, 1985). The study of drivers' judgemental processes identified a number of correlations with accident involvement. Excessive speed can be regarded as a judgemental error and several studies have identified excessive speed as a very im-portant contributory factor in accidents (e.g., Sabey & Staughton, 1975; Treat, 1980; Nilsson, 1981). Judgement of other road users' speed is also of relevance for safety in some traffic situations. Rumar and Berggrund (1973) found that subjects had difficulties in making correct judgements of distance and speed of an on-coming car when deciding whether to overtake. Hills and Johnson (1980) found that subjects tended to underestimate high speeds and overestimate low speeds. These findings lead to the conclusion that errors in information processing are dis-tinctly accident-related.

Information-induced errors

A basic problem is that much of the information which can be derived from the prevailing traffic situation is at best superfluous and at worst straightforwardly mis-leading; most is contradictory (e.g. Theeuwes & Godthelp, 1992). Providing addi-tional information, a task which the GIDS system is in principle capable of doing, does not necessarily help matters; first, because additional information may very well lead to further contradictory cues for the driver and, secondly, because the traffic situation is at times already so rich in information that supplying more of it would only serve to increase the overload of the driver. This last issue is discussed in more detail below.

Information-induced driver errors can be classified as follows:

(a) Insufficient information to select the appropriate action
In this situation the environment simply does not provide the information that the drivers require to adapt their behaviour to that situation. Among inexperienced

·drivers this may be due to the fact that the driver lacks the knowledge required to interpret the information (e.g., line markings that indicate where overtaking is not allowed). Among experienced drivers this may be due to the fact that the information is not salient·enough to change their reference (e.g., a gradual change from motorway to dual carriageway). In both cases, GIDS would have to provide additional information necessary for selecting the appropriate actions.

(b) Ambiguous information that leads to conflicting actions
In this type of situation drivers are not necessarily faced with a surplus of information, the information simply does not make sense or tells them to do different things at the same time. Navigation information (such as road signs) may direct drivers to a direction that is contradictory to their dead-reckoning knowledge, driving on a main road may require yielding to traffic on a seemingly secondary road, traffic lights may show red while there does not seem to be conflicting traffic, and so on. In such cases it is the task of the GIDS system to structure the incoming information hierarchically, such that the appropriate production is chosen regardless of the contradictory information.

(c) Surplus information that does not lead to any action
Some traffic situations supply too much of the right thing, that is, a surplus of information that cannot be meaningfully interpreted by the driver. With increasing density and complexity of traffic this type of situation is likely to become more frequent. It is already a major problem for the elderly who, for this reason, report that they tend to avoid situations which require rapid processing of different sources of information. In this case the support function obviously does not involve providing even more information; instead the GIDS system has to revert to explicit behavioural instructions. The issue of information overload is discussed in detail in the following section.

3.4 Driver workload

The concept of workload is central in cognitive ergonomics. For driver support workload is important in several ways. The GIDS system is to contain a device that monitors fluctuations in workload (i.e., a workload estimator) so as to intervene when momentary workload exceeds acceptable levels and to time its interventions such that it does not further increase workload which is already at a high level.

Workload and information processing

Early models of divided attention asserted that humans behave as a single channel with limited capacity for information processing and that they are unable to perform more than one thing at the same time. Later studies demonstrated little or no decrement in dual-task situations, however, and this led to the formulation of multi-capacity theories of human information processing. These theories state that human information processing depends on the allocation of separate resources to different processing stages so that humans can divide their attention efficiently between concurrent tasks provided that these tasks draw on separate rather than common processors or processing resources. The most influential of these multi-capacity theories is the multiple resource theory (Wickens, 1980, 1984) which states that tasks can be executed concurrently when the tasks utilize different modalities of input (e.g., visual versus auditory) and response (manual versus vocal), when they differ in the demands on certain stages of processing (perceptual, central, or motor processes), and when they require different codes of perceptual and central processing (spatial versus verbal codes). Even though recent research has indicated several shortcomings (e.g., Neumann, 1987; Verwey, 1991), multiple resource theory nevertheless appears to provide a practical framework for deciding how to measure workload in most driving situations and for indicating principles of interface design. This approach has been adopted in the development of GIDS.

Workload assessment methods

The concept of workload is used in several ways. Some researchers refer merely to perceptual-motor load, whereas others incorporate the task environment with its physical, social, and emotional components. In the latter case, the concept becomes almost equivalent to stress or strain. The term workload is also used to indicate load on the perceptual, central, and output resources (Wickens, 1980, 1984). In this case, 'mental' refers not only to workload on central resources but also incorporates workload on input and output modalities. In the GIDS context, workload will be interpreted in the latter, more comprehensive, sense. Mental workload has been operationalized in several ways, resulting in various methods of workload measurement. These methods yield different workload scores that are difficult to equate due to the absence of a theoretical foundation of workload assessment. In general, four methods of workload measurement can be distinguished: primary task performance, secondary task performance, physiological, and subjective measures.

Primary task performance is used as an indicator of workload under the assumption that an increase in task load reduces performance. Examples are error evaluation and reaction timing. Yet, if there is sufficient spare processing capacity an increase in workload will not necessarily be reflected in a task performance decrement. In that case it is necessary to establish the amount of spare processing capacity. Sanders (1979) proposed 'testing the limits' by increasing the difficulty of the primary task until performance decreases.

The *secondary task technique,* also referred to as the subsidiary task technique, assumes that an increase in workload on the primary task can be indicated by the level of performance on a second, lower priority task (Brown, 1964, 1978). The main task has to be performed optimally and variations in the workload of the primary task are indicated by variations in the performance on the secondary task (e.g., time interval production variability; Michon, 1966). However, the requirement, that performance on the primary task should not be affected by the secondary task, is often problematic (e.g., Noy, 1987). Secondary task performance can also be affected by interference with one or more resources used in common with the primary task, which makes performance on one particular secondary task difficult to interpret as a measure of general workload. Multiple resource theory postulates that the more similar a secondary task is to the primary task, the better indicator for general workload of the primary task it will be. However, in the case of a complex task such as driving, a battery of secondary tasks may be more informative (Kahneman, 1973); ideally the load on each resource would be indicated by performance on a different secondary task.

Physiological measures are also used in workload assessment. Some measures are primarily sensitive to workload upon specific processing resources (e.g., event-related potentials) whereas others (e.g., pupil diameter and heart rate variability) seem to index overall mental workload (Kramer, Sirevaag, & Braune, 1987; Roscoe, 1987; Wickens, 1984). Unfortunately, physiological measures are sensitive to artefacts such as physical workload, noise, and emotion-induced effects (Roscoe, 1987; Sanders, 1979). Also, their measurement is fairly intrusive and subject to large inter-individual differences. An advantage is that they are usually applicable without interfering directly with the task.

Finally, advocates of *subjective estimates or ratings* of workload assert that humans are quite capable of estimating task difficulty. Subjective estimates appear applicable in a wide range of tasks (e.g., Reid, 1985; Hart & Staveland, 1988). In particular when task demands are low, subjective estimates provide an accurate index of workload (Eggemeier & Stadler, 1984). When task demands exceed the capacity of working memory, subjective measures are less sensitive (Yeh & Wickens, 1988). Subjective estimates may not be obtainable during periods of very high workload due to the additional demands of making the load estimates, in particular if these are required on several dimensions (Roscoe, 1987).

Workload assessment in driving

The importance of workload in driving is intuitively apparent and research has demonstrated that, while driving a specific route, variations in mental load are related to variations in accident probability on that route (MacDonald, 1979). Here it is important to distinguish navigation, manoeuvring and vehicle control task levels as discussed in Section 3.2. The levels differ with respect to the load imposed on different processing resources (Wickens, 1984). Navigation in a relatively unfamiliar area mainly loads on central resources. Tasks at the control level of driving load on visual and motor resources rather than on central resources. In general, it can be concluded that tasks at a higher level of driving load more on central resources.

A number of empirical studies have been conducted to analyze the various workload assessment methods in driving. Wetherell (1981), for example, compared seven vocal-auditory tasks, secondary to the driving task, in order to measure load on central resources. The subjects' task was to travel a quiet rural road, which is basically a control task. No single secondary task could be identified as a general measure of the workload imposed by driving. Wetherell concluded that the secondary task most similar to the primary task is the most appropriate for measuring workload in driving, which is in line with recent theoretical work (e.g., Neumann, 1987; Schneider & Detweiler, 1988). Hicks and Wierwille (1979) compared five methods of measuring workload in a driving task in which gusts of wind applied to the front of the vehicle introduced varying levels of task difficulty. In this vehicle control task, subjective workload estimates and primary performance measures appeared the most sensitive to changes in task difficulty. Visual occlusion, cardiac arrhythmia, and secondary task performance were not significantly affected by variations in driving task difficulty.

Curry, Hieatt, and Wilde (1975) employed four techniques to compare mental workload assessment procedures in various driving conditions, one psychophysiological (heart rate and heart rate variability) and three secondary task techniques (random digit generation, time-interval production, and a short-term memory task). Cardiovascular measures and interval production measurement appeared the most sensitive measures. Biesta and Blaauw (1976) used heart rate, heart rate variability and an auditory detection task to assess driver workload resulting from static and dynamic aspects of the environment. They found heart rate and especially heart rate variability to be sensitive to static aspects of the road and traffic environment (e.g., the type of road). The auditory detection task appeared to be an insensitive measure of driver workload, while it affected driving performance and the heart rate measures, indicating that subjects had given too much attention to the secondary task, probably due to the obtrusiveness of stimulus presentation. The results may also indicate that auditory detection taps resources which are not used while driving, and would make auditory detection unsuitable for assessing driver workload. Janssen and Gaillard (1984) compared physiological measures (the 0.1 Hz component of the cardiac interval spectrum, and the P300 evoked response component of the EEG) and urinary catecholamine excretion rates in a route-choice experiment. All three measures appeared useful but the 0.1 Hz cardiac interval spectrum component was somewhat more sensitive and consistent. Finally, Verwey (1991) compared three secondary tasks to measure load on perceptual and central resources among inexperienced and experienced drivers. Performance on the secondary tasks was found to be a sensitive measure of workload and produced evidence for the notion that experienced drivers can be described best by a resource model with separate visual and central resources, whereas inexperienced drivers can be described best by a single resource model. Steering-action rate and mirror-glance frequency were found to be adequate indicators for load on visual and cognitive resources, except when mirror usage increases in task priority, as is the case during merging manoeuvres.

Workload and GIDS

From the above review it is clear that there is no single best method to assess driver workload. In evaluating interface designs, subjective estimates, steering-action rate and the 0.1 Hz cardiac component are sensitive measures for general workload assessment. Workload on particular resources can be determined by using a combination of secondary tasks. For the design of adaptive support systems, continuous workload assessment, utilizing multidimensional workload indices, can be employed. At present, workload estimations are most suitable, possibly in combination with on-line workload measurements.

Workload can be estimated by using a (descriptive) driver model in which the major factors determining driver workload are included. The workload estimator would give estimates of the load on perceptual, central, and possibly motor resources. In an initial attempt to find determinants of driver workload Verwey (1991) showed that workload is mainly determined by the traffic situation and to a lesser extent by driving experience. Further research is required to gain understanding of the factors determining workload among various groups of drivers. For the GIDS system, on-line workload measures can be used to tune workload estimates but are probably not sufficient. Workload measures can be adapted directly by increasing or decreasing the indices of load supplied by the workload estimator, or indirectly by altering the parameters in the driver workload model. In this respect, steering-action rate appears to be a useful measure as it is easy to obtain and a sensitive measure of general workload (Verwey, 1991). Finally, variability in driving speed and the occurrence of drivers' verbal and manual responses can also be used as indices of driver workload.

3.5 Conclusions

Adaptability is a central feature of intelligent driver support. Since drivers appear to differ considerably in their capabilities and habitual driving behaviour, intelligent support involves the provision of information as it is required by the driver. It appears very difficult to ascertain what exactly drivers need to enable them to drive safely. Young and inexperienced drivers have a high accident involvement, but the studies comparing the driving behaviour of this category of drivers with others with lower accident-involvement levels have not (yet) yielded conclusive results. It can be argued that these drivers will operate more on a knowledge-based or rule-based level and are therefore more prone to error due to inappropriate or inadequate problem-solving strategies. To prevent such errors these drivers would require support in the form of information about salient features in the traffic environment (to enable them to select the appropriate problem-solving strategy) as well as information about the actions required (to enable them to select an adequate problem-solving response). Experienced drivers, on the other hand, operate more on a skill-based level and are therefore more prone to errors due to monitoring failure, and would require support in that respect. At all levels, many errors are due to incorrect expectancy sets and GIDS support would involve alerting the driver to crucial changes in the momentary traffic situation. Elderly drivers seem to have an additional problem in complex or time-critical situations. Here GIDS support would essentially involve the reduction of workload by providing explicit instructions. Which type of support is appropriate is also dependent on the momentary workload faced by the driver. This depends, of course, largely on the driving situation, but is also driver dependent, partly in terms of stable characteristics (elderly) and partly in terms of momentary factors. An example of the latter is the navigation task of which the workload depends to a large extent on the familiarity of the environment. Adaptive support needs to take account of the momentary workload and on-line workload measurement is therefore essential. Steering-action rate appears a useful index for this purpose. Irrespective of the driver's characteristics, feedback seems to be a crucial factor in shaping appropriate productions. As drivers appear difficult to categorize and insufficient empirical data are available to do so with precision, adaptation to the driver's personal characteristics will be essential to reach optimal adaptivity. This personalized support is discussed further in the following chapter (Chapter 4).

References

Avolio, B.J., Kroeck, K.G., & Panek, P.E. (1985). Individual differences in information-processing ability as a predictor of motor vehicle accidents. *Human Factors, 27*, 577-587.

Biesta, P.W., & Blaauw, G.J. (1976). *Effecten van omgevingsfactoren op maten voor werkbelasting en rijgedrag* [The effects of environmental factors on measures for workload and driving behaviour]. Report IZF 1976-6. Soesterberg. The Netherlands: TNO Institute for Perception.

Brendemühl, D., Schmidt, U., & Schenk, N. (1988). Driving behaviour of elderly motorists in standardized test runs under road traffic conditions. In J.A. Rothengatter & R.A. de Bruin (Eds.), *Road user behaviour: Theory and research* (pp. 310-318). Assen, The Netherlands: Van Gorcum.

Brookhuis, K.A. (1990). *DREAM - Second annual review report.* Report to the Commission of European Communities. Haren, The Netherlands: Traffic Research Centre, University of Groningen.

Brown, I.D. (1964). The measurement of perceptual load and reserve capacity. *Transactions of the Association of Industrial Medical Officers, 14,* 44-49.

Brown, I.D. (1978). Dual-task methods of assessing workload. *Ergonomics, 21,* 221-224.

Brown, I.D. (1982). Exposure and experience are a confounded nuisance in research on driver behaviour. *Accident Analysis and Prevention, 14,* 345-352.

Curry, G.A., Hieatt, D.J., & Wilde, G.J.S. (1975). *Task load in the motor vehicle operator: A comparative study of assessment procedures.* Report CR7504. Ottawa: Ministry of Transport.

Duncan, J, Williams, P., & Brown, I. (1991). Components of driving skill: Experience does not mean expertise. *Ergonomics, 34, 919-937.*

Eggemeier, F.T., & Stadler, M.A. (1984). Subjective workload assessment in a spatial memory task. *Proceedings of the Human Factors Society, 29,* 680-684.

Engels, K., & Dellen, R.G. (1983). *Beitrag zur Quantifizierung des Altersrisiko von PKW-Fahrern.* Unfall- und Sicherheitsforschung Strassenverkehr, Heft 42. Köln, Germany: Bundesanstalt für Strassenwesen.

Evans, L. (1991). *Traffic safety and the driver.* New York: Van Nostrand Reinhold.

Finn, P., & Bragg, B.W.E., (1986). Perception of the risk of an accident by younger and older drivers. *Accident Analysis and Prevention, 18,* 289-298.

Fontaine, H. (1988). Wie krijgt wettelijk de schuld bij verkeersongevallen? [Who is legally blamed for traffic accidents?]. *Verkeerskunde, 39,* 101-103.

Groeger, J.A. (1991). Supporting training drivers and the prospects for later learning. In *Proceedings of the CEC DRIVE Conference Advanced Telematics in Transport* (Vol. 1, pp. 314-330). Amsterdam: Elsevier.

Groeger, J.A., & Brown, I.D. (1989). Assessing one's own and others' driving ability: Influences of sex, age and experience. *Accident Analysis and Prevention, 21,* 155-168.

Groeger, J.A., Kuiken, M.J., Grande, G., Miltenburg, P.G.M., Brown, I.D., & Rothengatter, J.A. (1990). *Preliminary design specifications for appropriate feedback provision to drivers with differing levels of traffic experience.* Deliverable Report DRIVE V1041/GIDS-ADA 01. Haren, The Netherlands: Traffic Research Centre, University of Groningen.

Hart, S.G.. & Staveland, L.E. (1988). Development of NASA-TLX (Task Load Index): Results of empirical and theoretical research. In P.A. Hancock and N. Meshkati (Eds.), *Human mental workload* (pp. 139-183). Amsterdam: North-Holland.

Hicks, T.G., & Wierwille, W.W. (1979). Comparison of five mental workload assessment procedures in a moving-base simulator. *Human Factors, 21,* 129-143.

Hills, B.L. (1980). Vision, visibility, and perception in driving. *Perception, 9*, 183-216.

Hills, B.L., & Johnson, L. (1980). Speed and minimum gap acceptance judgements at two rural junctions. Reference in B.L. Hills, Vision, visibility, and perception in driving. *Perception, 9,* 183-216.

Janssen, W.H., & Gaillard, A.W.K. (1984). *Task load and stress on the road: Preliminaries to a model of route choice.* IZF Report 1984 - C-10. Soesterberg, The Netherlands: TNO Institute for Perception.

Kahneman, D. (1973). *Attention and effort.* Englewood Cliffs, NJ: Prentice-Hall.

Kahneman, D., Ben-Ishai, R., & Lotan, M. (1973). Relation of a test of attention to road accidents. *Journal of Applied Psychology, 58*, 113-115.

Kramer, A.F., Sirevaag, E.J., & Braune, R. (1987). A psychophysiological assessment of operator workload during simulated flight missions. *Human Factors, 29*, 145-160.

Kuiken, M.J., & Groeger, J.A. (Eds.). (1991). *Report on feedback requirements and performance differences.* Deliverable Report DRIVE V1041/GIDS-ADA 02. Haren, The Netherlands: Traffic Research Centre, University of Groningen.

MacDonald, W.A. (1979). *The measurement of driving task demand.* Melbourne: Corvina.

Matthews, M.L., & Moran, A.R. (1986). Age differences in male drivers' perception of accident risk: The role of perceived driving ability. *Accident Analysis and Prevention, 18*, 299-313.

Maycock, G., Lockwood, C.R., & Lester, J.F. (1991). *The accident liability of drivers.* TRRL Research Report 315. Crowthorne, United Kingdom: Transport and Road Research Laboratory.

McKnight, A.J., & Adams, B.B. (1970). *Driver education task analysis. Volume I: Task descriptions.* Final Report, contract no. FH 11-7336. Alexandria, VA: Human Resources Research Organization.

Michon, J.A. (1966). Tapping regularity as a measure of perceptual-motor load. *Ergonomics, 9*, 401-412.

Michon, J.A. (1985). A critical view of driver behavior models: What do we know, what should we do? In L.A. Evans and R.C. Schwing (Eds.), *Human behavior and traffic safety* (pp. 487-525). New York: Plenum Press.

Michon, J.A., Smiley, A., & Aasman, J. (1990). Errors and driver support systems. *Ergonomics, 33*, 1215-1229.

Miltenburg, P.G.M., & Kuiken, M.J. (1991). Driving performance and observational strategies of novice and experienced drivers. In Y. Quéinnec and F. Daniellou (Eds.), *Designing for everyone. Proceedings of the 11th Congress of the International Ergonomics Association* (pp. 1544-1546). London: Taylor and Francis.

Näätänen, R.. & Summala, H. (1974). A model for the role of motivational factors in drivers' decision making. *Accident Analysis and Prevention, 6*, 243-261.

Näätänen, R., & Summala, H. (1976). *Road-user behavior and traffic accidents.* Amsterdam: North-Holland/American Elsevier.

Neumann, O. (1987). Beyond capacity: A functional view of attention. In H. Heuer and A.F. Sanders (Eds.), *Perspectives on perception and action* (pp. 361-394). Hillsdale, NJ: Erlbaum.

Nilsson, G. (1981). The effects of speed limits on traffic accidents in Sweden. In *Proceedings of the OECD Symposium on the effects of speed limits on traffic accidents and transport use* (pp. 1-9). Dublin: An Foras Forbatha.

Nilsson, L., & Alm, H. (1991). *Effects of mobile telephone use on elderly drivers' behaviour - including comparisons to young drivers' behaviour.* Report 176. Linköping, Sweden: Swedish Road and Traffic Research Institute.

Noy, Y.I. (1987). Theoretical review of the secondary task methodology for evaluating intelligent automobile displays. *Proceedings of the 31th Human Factors Society* (pp. 205-209). Santa Monica, CA: Human Factors Society.

O'Hanlon, J.F. (1981). Boredom: Practical consequences and a theory. *Acta Psychologica, 49,* 53-82.

Pelz, D.C., & Schuman, S.H. (1971). Are young drivers really more dangerous after controlling for exposure and experience? *Journal of Safety Research, 3,* 68-79.

Quenault, S.W., & Parker P.M. (1973). *Driver behaviour: Newly-qualified drivers.* TRRL Report LR567. Crowthorne, United Kingdom: Transport and Road Research Laboratory.

Reid, G.B. (1985). Current status of the development of the subjective workload assessment technique. *Proceedings of the Human Factors Society, 29,* 220-223.

Roscoe, A.H. (1987). Introduction. In A.H. Roscoe (Ed.), *The practical assessment of pilot workload* (pp. 1-10). Neuilly-sur-Seine, France: AGARD.

Rothengatter, J.A. (Ed.) (1989). *Navigation information requirements: literature review.* Deliverable Report DRIVE V1041/GIDS-NAV 01. Haren, The Netherlands: Traffic Research Centre, University of Groningen.

Rothengatter, J.A. (1991). Agressie in het verkeer [Aggression in road traffic]. In P.B. Defares & J.D. van der Ploeg (Eds.), *Agressie: Determinanten, signalering en interventie* (pp. 68-75). Assen, The Netherlands: Van Gorcum.

Rothengatter, J.A., & Harper, J. (1991). The scope and design of automatic policing information systems with limited artificial intelligence. In *Proceedings of the CEC DRIVE Conference on Advanced Telematics in Transport* (Vol. 2, pp. 1499-1515). Amsterdam: Elsevier.

Rumar, K. (1990). The basic driver error: Late detection. *Ergonomics, 33,* 1281-1290.

Rumar, K., & Berggrund, U. (1973). Overtaking performance under controlled conditions. *Paper presented at the First International Conference on Driver Behaviour* (IDBRA), Zürich.

Sabey, B.E., & Staughton, G.C. (1975). Interacting roles of road environment, vehicle and road user in accidents. *Paper presented to the 5th international conference of the International Association for Accidents and Traffic Medicine*, London.

Sanders, A.F. (1979). Some remarks on mental load. In N. Moray (Ed.), *Mental workload: Its theory and measurement* (pp. 41-77). New York: Plenum Press.

Schank, R.C., & Abelson, R.P. (1977). Scripts, plans and knowledge. In P.N. Johnson-Laird and P.C. Wason (Eds.), *Thinking* (pp. 421-432). Cambridge: Cambridge University Press.

Schneider, W., & Detweiler, M. (1988). The role of practice in dual-task performance: Towards workload modeling in a connectionist/control architecture. *Human Factors, 30,* 539-566.

Smiley, A., & Brookhuis, K.A. (1987). Alcohol, drugs and traffic safety. In J.A. Rothengatter and R.A. de Bruin (Eds.), *Road users and traffic safety* (pp. 83-104). Assen, The Netherlands: Van Gorcum.

Svenson, O. (1981). Are we all less risky and more skillful than our fellow drivers? *Acta Psychologica, 47,* 119-133.

Theeuwes, J., & Godthelp, J. (1992). De begrijpelijkheid van de weg. [The comprehensibility of the road]. Report IZF-1992-C8. Soesterberg, The Netherlands: TNO Institute for Perception.

Treat, J.R. (1980). *A study of precrash factors involved in traffic accidents.* Highway Safety Research Institute Research Review. HSRI 10/11, 6/1. Ann Arbor, MI: Highway Safety Research Institute.

Van Kampen, L.T.B. (1988). *Analyse van de verkeersonveiligheid van jonge, onervaren automobilisten* [Analysis of the traffic risk of young, inexperienced drivers]. Leidschendam, The Netherlands: Institute for Road Safety Research SWOV.

Van Wolffelaar, P.C. (1988). *Oudere verkeersdeelnemers: Verkeersproblemen en educatiedoelstellingen* [The elderly in traffic: problem analysis and educational objectives]. Report VK88-16. Haren, The Netherlands: Traffic Research Centre, University of Groningen.

Van Wolffelaar, P.C., Brouwer, W.H., & Rothengatter, J.A. (1990). *Divided attention in RTI-tasks for elderly drivers.* Report 1006/TRC1 to the Commission of European Communities. Haren, The Netherlands: Traffic Research Centre, University of Groningen.

Verwey, W.B. (1991). Towards guidelines for in-car information management: Driver workload in specific driving situations. Report IZF 1991-C 13. Soesterberg, The Netherlands: TNO Institute for Perception.

Warnes, A.M., Frazer, D., & Rothengatter, J.A. (1991). Elderly drivers' reactions to new vehicle information systems. In *Proceedings of the CEC Drive Conference Advanced Telematics in Transport* (Vol. 1, pp. 331-350). Amsterdam: Elsevier.

Welford, A.T. (1982). Motor skills and aging. In J.A. Mortimer, F.J. Pirozzolo, & G.J. Maletta (Eds.), *The aging motor system* (pp. 152-187). New York: Praeger.

Wetherell, A. (1981). The efficacy of some audio-vocal subsidiary tasks as measure of the mental load on male and female drivers. *Ergonomics, 24,* 197-214.

Wickens, C.D. (1980). The structure of attentional resources. In R. Nickerson (Ed.), *Attention and Performance VIII* . Hillsdale, NJ: Erlbaum.

Wickens, C.D. (1984). *Engineering psychology and human performance.* Columbus, OH: Merill.

Yeh, Y.Y., & Wickens, C.D. (1988). Dissociation of performance and subjective measures of workload. *Human Factors, 30,* 111-120.

Chapter 4
Driver support

Wiel H. Janssen, Håkan Alm, John A. Michon, Alison Smiley

4.0 Chapter outline

In this chapter driver support is considered from a conceptual point of view. This extends and expands what has been said about driving and the driving task in the preceding chapters, taking into account some insights gained from support systems that have been developed for other means of transport. In Section 4.1 a theoretical basis is provided. In the next Sections driver support is looked at from the applied, ergonomic point of view in a discussion of the actual constraints that should be met by a feasible driver support system, given the present state-of-the-art in road vehicle technology. These sections cover navigation (Section 4.2), manoeuvring (Section 4.3), control (Section 4.4), instruction (Section 4.5), and collateral support (Section 4.6), respectively.

4.1 The concept of driver support

Conceptual levels of information

As soon as one begins thinking of the potential of driver support, one is overwhelmed by the sheer number of possibilities. Even worse, every extra feature added by some information source to the driving scenario under consideration would, theoretically, double the time required to evaluate the consequences of the extra feature in relation to the information that was already available. Clearly, therefore, we need to constrain and organize this dangerously explosive situation.

People have always wanted to extend the power and reach of their senses and have always looked for support. Modern drivers do not differ in this respect from their hunting and gathering ancestors. Actually the nature of the problems involved has remained pretty much the same, because our senses and our cognitive capabilities have not really changed much since prehistoric times: we are born with capabilities that are geared to the environmental conditions of the Tertiary era. Of course the environment has changed. We must therefore take a look at the present as well. The development of cognitive science has brought us the computational approach to human information processing (Stillings et al., 1987; Posner, 1989). This approach to natural and artificial intelligence and communication allows us to establish a useful, theory-based frame of reference for the design and development of driver support systems.

Altogether a distinction may be made between four kinds of driver support, ways of catering to the driver's information needs. These four kinds are reflected in theoretical concepts that, in principle, enable the analysis and formalization of the relevant issues.

The first type includes such items as road markings that provide perceptual *enhancement,* that is, they support people by allowing them to see more clearly what is already there but difficult for everyone to see. We may say that this type of support influences the *amount* of information available.

The second type of support, *augmentation,* is somewhat more sophisticated. Augmentation includes, among other things, such items as dynamic road signs that warn us of impending hazards. That is, they extend the range of cues so as to facilitate the identification of the traffic situation and to reduce the uncertainty. As such they can be said to affect the *form* of the information available to the traveller.

The third type constitutes an altogether different category, including in particular the so called 'on-board computers' that provide information about the status of the road or one's car. Their principal function is the *interpretation* of inputs from the environment within a fixed, preset context. Any adaptive use outside the predetermined context must come from the user. This type of system affects the *meaning* of the information presented.

The fourth and final type introduces adaptive features. It differs from the previous type in that systems in this category are not rigidly programmed but, instead faithfully follow their user's intentions and goals (although on occasion they may, of course, misunderstand their user). Their function is to *support* their masters, filtering, integrating, and presenting information in ways that are consistent with the user's goals and present information needs. Consequently they may be said to affect the *purpose* or the pragmatic aspect of information.

To summarize, drivers can be provided with additional information about the situation in which they are performing their task in four distinct ways: by means of enhancement, augmentation, interpretation, and support (see Table 4.1). These four can be identified theoretically as the *metric, syntactic, semantic,* and *pragmatic* aspects of human information processing, respectively.

Table 4.1 Four levels of information processing

Function	Quality	Theoretical Level
Enhancement	Amount	Metric
Augmentation	Form	Syntax
Interpretation	Meaning	Semantic
Support	Purpose (Goal)	Pragmatic

Support functions

The central function of systems of the fourth kind, that is, of adaptive driver support systems such as GIDS, is to organize the information that is to be processed by the vehicle operator, by filtering, interpreting, integrating, prioritizing, and presenting the information from any number of sensor systems and applications in the vehicle and the environment. The unconstrained availability of microelectronics and telecommunication applications is vastly increasing the potential workload on the driver, but at the same time offers an unprecedented opportunity for sophisticated driver support. We should therefore be able to provide functionality of all four kinds mentioned above. Table 4.2 extends the classification given above. Observe that the listed functions are given in an increasing order of adaptive control by the co-driver.

Table 4.2 Nine types of basic driver support functions that may be implemented in an intelligent driver support system, in increasing order of adaptive control exercised by the system. Examples are given in brackets.

- enhancing information (increasing visibility by retroflection)
- augmentation (special information about icy patches)
- warning (against speeding or other violations)
- advice (to take a less congested route)
- explanation (reason for delay, e.g., accident ahead)
- instruction (feedback about incorrect action)
- intervention (speed delimiter)
- substitute or secondary control (cooperative driving)
- autonomous or primary control (robot driving)

GIDS will offer warnings, advice, and instruction, but it will, generally speaking, not intervene or take control. However, we shall see that we need to allow one important exception. Some elementary control actions, such as steering and braking, require a very rapid response from the driver. As we shall see later, in these cases GIDS will provide support that is close to cooperative driving

Driver support systems derive their usefulness to a large extent from the fact that vehicle operators must cope with a growing amount of information of an increasingly complicated nature. This is caused by several factors, including the increasing traffic intensity, an increasing number of on-board and roadside sources of information and, not least, by the increasing amount of additional in-vehicle equipment, such as telephones and fax machines. This avalanche of information – much of which may eventually be generated by RTI systems resulting from the DRIVE programme – is likely to have an impact on almost every aspect of the driving task. It will affect route planning as well as navigation, manoeuvring, or elementary vehicle control. Unless regulatory action is taken this information will eventually be presented to the driver in an essentially incoherent fashion, irrespective of meaning or urgency. An important function of GIDS, as of any other co-driver, is to protect the driver from such uncoordinated information.

The really innovative feature of GIDS is that it is the first system ever to take into account, in an adaptive fashion, the intentions, capabilities, and limitations of the individual driver. For this purpose GIDS incorporates not only a knowledge base containing detailed scenarios for driving manoeuvres such as overtaking, negotiating intersections, or merging into a traffic stream, but also a knowledge base containing information about the characteristics and behavioural history of individual drivers.

The components of an intelligent support system

As pointed out already in Chapter 1, an intelligent driver support system, such as GIDS, must have at least four components that contain the required knowledge about the world in which the system is supposed to operate: it must have internal representations of (a) the driving task and the environment; (b) the driver; (c) the nature of any discrepancy between required and prevailing situation; and (d) appropriate actions for warning, advising, or instructing the driver. Let us consider these four elements in turn (see also Figure 4.1).

(a) The driving task
Knowledge about the driving task is stored in a way that we may conceptualize as an idealized model driver. This *reference driver* should have the same limitations as a typical, experienced, healthy driver in terms of basic perfomance parameters, such as reaction times or visual acuity. The resemblance requirement does not necessarily include variables that relate to driving style. A typical driver may, for instance, drive faster at night than is safe, or may enter curves too quickly, or follow other cars too closely. The characteristics of the reference driver most useful to a GIDS system would be a blend of 'ideal' – or 'normative' – and typical driving characteristics. If only typical characteristics were to be contained in the knowledge base for the reference driver, novice drivers would be tutored so that they would more quickly attain the bad habits of experienced drivers. If only ideal characteristics were to be used, then the reaction time to a warning signal expected for some drivers would become unrealistically short, and therefore unsafe. Thus the

reference driver should represent at the same time 'good' driving performance, to the best of our ability to define it formally, as well as the realistic range of expectations and habits of actual drivers.

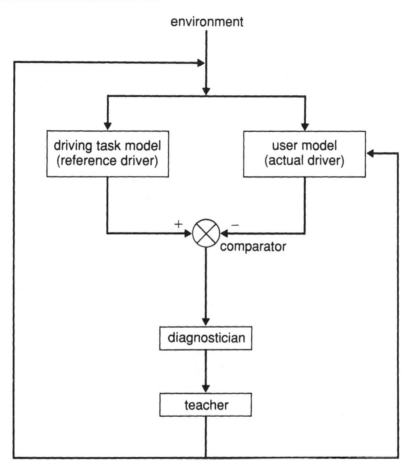

Figure 4.1 Basic components and information flow of an intelligent support system

In Chapter 3 we have seen that what constitutes 'good' driving is most certainly a contentious issue. First of all, we have to acknowledge that there is a dearth of information about 'normal' driving by subjects of various ages, levels of experience, cultures, and abilities. Whilst it is certainly possible to specify normal ranges of, for example, speed variability for female university students, no such definite characteristics can be given for the population as a whole. Second, even if we had these data, it would be difficult to define what is 'good', except in a rather arbitrary fashion. Our standards for 'good' tend to depend on the state of the driver, rather than on any objective criterion. For example, a lane position variability of 0.5 m might be attained by sober drivers of a few years' experience, in which case we would qualify such performance as 'good'. However, the same lane position variability obtained by very experienced, but slightly drunk, drivers would be considered 'bad' simply because it reflects poorer performance than would be possible if these drivers were sober. Third, driving is a combination of tasks. A lane position variability of 0.5 m would not be considered good if it were to be attained at the cost of reduced ability to monitor other traffic. Governments might at some future point require that this reference driver be not only a 'good' driver but also a legal one. If that is the case, warnings would be given when the driver exceeds the speed limit, or did not stop before the white line, the way some cars warn drivers that their seat-belts are unfastened.

To be able to perform the driving task efficiently and safely, the reference driver must also have knowledge of the driving task, the particular vehicle being driven and the road traffic environment. In terms of the vehicle being driven, the system would need to know the kind of braking system, the weight of the vehicle (a trailer, for instance, will change braking capacity), the kind of tyres, the condition of brakes and tyres, and so on to determine braking distances at various speeds, on various road types, under various road conditions (e.g., gravel in the rain). These braking distances will influence at what distances from a car in front warnings are given. Other aspects of the vehicle being driven should also be considered. For example, the navigation system should know that the vehicle being driven is a truck of a particular height and width and weight, to avoid directing the driver along roads prohibited to such vehicles.

Knowledge of the road traffic environment should include such items as expected stopping distances of other vehicles (for example, as the driver of the GIDS-equipped vehicle approaches a red light behind another car), coefficients of friction of various road surfaces and the effects of various weather conditions on these (which will then help determine expected stopping distances).

(b) The actual driver

For a support system to be intelligent it also requires knowledge of the driver who is, currently, operating the vehicle. The system must therefore contain a representation of the particular characteristics, both long term and short term, of that driver, in short, it needs a model of the actual driver. Particular characteristics relate both to level of experience and individual idiosyncrasies. Examples of the latter are very slow reaction times, or habitually cautious braking. An intelligent system will present its warnings to such a driver earlier than it would to the reference driver. If the driver is a novice these characteristics will change fairly rapidly over time and the actual driver model needs continual updating to accurately represent the driver.

The actual driver model should also accurately represent short-term changes in behaviour, such as a reduction in alertness (assuming for the sake of argument that this might be reliably concluded from the absence of steering wheel movements for the preceding 10 seconds), or an increase in aggressive behaviour (assuming that this is revealed by repeated rapid acceleration or by keeping very short headways). Since the characteristics of a particular driver change over time, the actual driver model must be updatable, with permanent and temporary changes. How the actual driver representation will be updated and which updates should be temporary and which permanent, are difficult matters that require much consideration.

The above description of the actual driver model considers how the driver indirectly influences the actual driver model. The driver should also be able to influence the actual driver model through direct inputs. For example, a driver might directly indicate his/her level of subjective mental load to the system. If too high, an intelligent system might respond by reducing the information presented and increasing the length of each presentation.

Initially, before the system has any experience with the actual driver there would need to be some standard settings assumed for braking forces, reaction times, acceleration patterns and so on. Over time, as the system gains knowledge of a particular driver these standard settings would become modified until they more closely represented the actual driver. How many such standard settings there should be is a question for further discussion. For example, one might have different standard settings for young, middle-aged, and old drivers, for novice and experienced drivers, for drivers with particular handicaps and so on.

(c) Discrepancy between desired and actual behaviour (the diagnostician)
The next conceptual component in a driver support system should be able to compare characteristics of the behaviour of the actual driver with those of the reference driver, and determine the appropriate action to be taken on the basis of this analysis. It is by this diagnostic, or analytic, means that a driver support system can perform in an adaptive manner. Thus it may, for example, take automatic action where an inexperienced driver might take too long to respond, or setting the warning level for obstacles at a lower threshold for a driver who was consistently slow to react to such warnings. The system could also act in a tutorial manner for an inexperienced driver, for example, by giving feedback about the adequacy of gaps selected for passing at the start of the overtaking manoeuvre.

In order to determine the appropriate support to offer the driver, the analyst should contain knowledge of the driving support task. It should be noted that this is different from the knowledge of the driving task contained by the reference driver component. For example, a driver who sees a car in front uses angle-subtended-at-the-eye cues to tell him how fast the car is moving relative to himself. If the driver detects that he is approaching he will slow down, but the degree of slowing down will not be correlated very well with the relative closing velocity, because the driver is so poor at perceiving this. The GIDS support task, on the other hand, involves different cues, namely radar signals, to determine closing speed. The support task involves giving the driver very precise information about the closing speed and a warning if the closing speed is high enough that a collision could result.

Each support system needs a sub-analyst component to determine required actions for that particular task. Within each module the selection of messages should be determined according to the current situation with regard to the task in which the driver is being supported. Thus the obstacle detection analyst module, once it detects too high a closing speed to the car in front, would select a message to that effect to be passed to the dialogue controller. The need for a warning should be determined based on information about the driver (e.g., drowsy, with a habitually slow reaction time), information about the car (e.g., towing a trailer, and brake systems in poor condition), information about the environment (e.g., slippery road due to rain) and information about the current driver task (e.g., approaching a caution light where the driver in front is likely to stop suddenly).

The analyst module should not only select the message, it should give it a priority, in this case urgent. Thus each support system would pass messages with priority levels encoded to the dialogue controller. Messages could also be encoded at this stage with various directions to the dialogue controller. These might include the time within which the message must be presented in order to be of use to the driver, and the amount of time the message must be displayed so that the driver can process it.

(d) The dialogue controller (the teacher)
The dialogue controller should integrate the outputs from the individual support modules and again select and prioritize information to be presented to the driver. The dialogue controller would need a model of the driving task to enable it to determine which of the equal priority messages from the three support modules has ultimate priority, and the appropriate order and timing of presentation. If the driver is negotiating a bend, for example, the system should not display even urgent messages from the navigation system until the driver is going straight again. If an information display has to be interrupted by a more urgent message, then the dialogue controller should re-initiate the display of the original message at an appropriate time.

Determining the priorities of the various messages is by no means a trivial task. One cannot assume for example that obstacle detection messages always take precedence over navigation messages. This is because, within each subtask, messages will have different priorities; some will offer the driver information (e.g., there is a car in front, you will encounter rain during the next part of your trip), some will offer the driver advice (e.g., you are following too close to the car in front, the road ahead is congested, try an alternative route) and some will instruct the driver (e.g., brake immediately, turn left, the road ahead is closed). An urgent instruction from the navigation system may take precedence over information from the obstacle detection system that there is a car ahead. Under foggy conditions, however, the priorities might be reversed.

Just as the driver should be able to make direct inputs to the actual driver model, he should be able to do the same with the dialogue controller. A driver might, for example, indicate that he has a high level of familiarity with a particular area, so that the navigational support would be reduced to presenting only information on recent changes such as closed roads or construction zones. A driver should be able, if he wishes, to turn individual support systems off.

4.2 Navigation

The purpose of all vehicle travel, perhaps with the exception of some pleasure trips, is to proceed from a known origin to a given destination. It is the navigation sub-task of the driving task that determines that one actually gets there.

Sometimes travelling occurs in a partly or totally unfamiliar environment. In such cases, many drivers could use the support of a navigation system which told them how to reach their destination, and provided important attributes connected with alternative routes. After the choice of a particular route the system would moreover guide drivers to their destination. If something unexpected happened during the trip, for instance an accident somewhere on the route, the system could revise the current plan and suggest another route. And, of course, the system would explain why a revision was necessary. When the driver had arrived at the destination the system would, finally, be able to give information about different types of parking opportunities and the costs associated with them.

From a traffic safety point of view two aspects of navigational support are particularly important.

First, it is crucial that the system does not distract the driver. Since driving to a large degree is dominated by visual information (Rockwell,1972) it seems wise to avoid introducing more visual information. If visual information must be used then we should strive to make the information as simple as possible. According to Zwahlen et al. (1988) an in-car system should not demand more than three fixations, each less than three seconds. After this time period the driver should be able to understand and also, if needed, remember the message from an in-car system. This means, probably in all cases, that a map-based navigation system should not be used when the car is moving. It also means that sense modalities other than the visual should, if possible, be used. One very good candidate for navigational guidance is verbally based information, which has some advantages over visual information (Streeter et al.,1985; Verwey & Janssen, 1988; Alm et al., 1991).

Second, the messages provided by the system should not lead to a dangerous increase in the driver's mental workload. An increased workload may interfere with the processing of driving related information, if drivers focus their attention on the information from the system. Conversely, if attention were to remain focused on the driving task, information from the system might be ignored.

To avoid the risks connected with a high level of workload the system should be able to 'speak the same language' as the driver. When the system gives reference to the driving environment it should do so in the same way as the driver thinks about the environment. The output from the system should closely match the driver's mental model of the driving environment. This will ensure that the driver easily grasps the messages from the system. The system should also provide drivers with information that reduces anxiety about being lost, and which reduces suspicion that the system may have stopped working.

Since drivers will differ in their knowledge of a particular driving environment it is necessary that a GIDS navigation system adapts to the individual's level of knowledge. Drivers who are totally unfamiliar with a driving environment may need directions guiding them from choice point to choice point. Drivers who are familiar with an environment may only need information telling them that route X is blocked, for a specified reason.

4.3 Manoeuvring: collision avoidance

Another essential subtask in driving is dealing with other vehicles on the road. This is commonly termed the manoeuvring level of the driving task. Collisions with other vehicles are pertinent evidence that this subtask is not always performed flawlessly.

A collision avoidance system (CAS), correcting faulty user perception or decision making, could reduce the frequency of collisions. Thus, a CAS could well become part of the in-car environment and thereby contribute to the stream of information directed towards the driver. This has been the motive for the study of a CAS as a component in a future GIDS system.

Although the technology for detecting the presence of obstacles and estimating their parameters of movement is by no means perfect it may be assumed that it will become so within a reasonable number of years. The question then becomes in what form this information should be put in order to be of use to the average, and possibly the not-so-average, driver.

A feeling for the purpose of a CAS may be obtained by noting that the average driver, at least in highly motorized countries, has a collision involving property damage every 10 or 20 years. Ideally a CAS should alert the driver in that case and only that one. A few more cases may be added if narrow escapes are included. Even then, however, it will be clear that superb discriminative power will be required from a CAS. Such power is probably unattainable, even if the detection of obstacles and their parameters of movement were perfect, because it would demand complete knowledge of what distinguishes collision from non-collision configurations well before the collision happened. Thus the system would not only have to recognize at a sufficiently early stage that a collision would follow if no action were taken, it would also have to know that the driver was not going to take evasive action in precisely that type of configuration.The dilemma obviously is that waiting longer, in order to make certain that there is in fact a critical situation, reduces the chances of taking action that will be effective in avoiding the collision, whilst signalling at a very early stage will produce large numbers of false alarms.

There is no easy way out of this dilemma, and recommendations for the best design of a CAS are less derivable from already existing results than for the navigational component. However, it might be wise not to look at collision-avoiding performance per se, but to evaluate a candidate CAS in terms of its general effects on behaviour. For example, it may be a better question to ask of a CAS how much it succeeds in bringing down people's tendency to follow preceding vehicles at very close headways instead of asking how many collisions it will avoid. This is in fact the approach that has been taken in GIDS, and that formed the basis for the research performed in that context. Chapter 7 deals further with this issue, as well as with the research itself that has been performed.

4.4 Control

The aim of the third subtask of the driving task is, generally speaking, to keep the vehicle on the appropriate part of the road. Single-vehicle accidents in which no other road user is involved are typically the result of control task errors.

We distinguish two kinds of control behaviour, lateral and longitudinal. Control theory further differentiates between 'closed loop' and 'open loop' control (McRuer et al., 1977). In closed loop control, feedback information is used continuously and the driver is in an error correcting mode. By contrast, open loop control is performed without feedback information. This control mode is executed with special motor programs, which operate on the perceived position of the car and the driver's internal model of the dynamic characteristics of the car.

If we look at the actions that are required in conventional driving we note that some are a combination of open and closed loop behaviour, like steering, whilst others are mainly open loop (overtaking) or closed loop (changing lanes). Looking at the way control actions are actually performed it is clear that lateral control is done with the steering wheel, and longitudinal control with the accelerator pedal. However, these two control modes are not totally independent. For instance, if a driver enters a bend at too high a speed he will leave the road even if his steering behaviour is perfect. Thus, optimizing lateral or longitudinal control separately is insufficient for perfect driving: there must be an appropriate relationship between the two. The intelligent GIDS co-driver is meant to find exactly that equilibrium under varying circumstances.

The design of the 'control' component of GIDS should realize that the visual attention of drivers should not be distracted from the road at a time when they have to concentrate on the environment in order to optimize their steering or speed control. Sheridan and Ferrell (1974) argue in their theory of perception-action compatibility, that feedback information should be presented at the actuators of control themselves. These are the hand/arm system and the feet of the drivers. Shorter reaction times and more appropriate reactions are expected when the relevant information for longitudinal control comes via the accelerator and for lateral control via the steering wheel. This kind of feedback is called tactual (Schiff and Foulke, 1982) or proprioceptive-tactual (Rühmann, 1981).

Given the decision to warn the driver via accelerator and steering wheel we have a problem: what should the warning feel like in order to produce the desired reaction? Short or long vibrations, constant counterforces and many other signals are conceivable. Some first steps toward solving the problem, by developing a tactual-proprioceptive logic that will be interpreted appropriately by the driver, are described in Chapter 7.

4.5 Behavioural feedback and instruction

A GIDS-system must must be able to adapt to individual differences in states and traits. In particular, attention should be given to the specification and evaluation of driver support which would meet the different needs of drivers with differing levels of knowledge and experience of the traffic system.

The complete GIDS-system will therefore have to include a means of storing the performance profiles of individual drivers, both in terms of the frequency with which they have encountered particular situations and their history of 'abnormal' performance in each situation. Abnormal performance will be regarded as a deviation from a normative standard in terms of the functioning of each sensor/application, such as sudden harsh braking, attempts to make rapid and unusually large steering corrections, excessive accelerations and adoption of highly variable headways. Initally, estimated performance criteria, in terms of acceptable numbers of incidences of such abnormal performance will be stored along with the performance profile for each situation. Exceeding this criterion will result in a request being passed to the Dialogue Controller (which schedules all communications to the driver) for a message to be presented to the driver, indicating that over a period of time there has been a tendency for (specified) deviant performance to occur in a particular situation, for instance a tendency for that driver to require excessive steering corrections while taking the third exit from a four-legged roundabout, together with advice about how to correct the error.

An 'individual feedback' support module in a GIDS system could be of particular relevance to novice drivers. Behavioural feedback is currently provided to trainees by professional driving instructors throughout a normal course of driving instruction, undertaken by individuals who are completely naive with respect to driving. Driving instructors supposedly adapt the level of support they provide to the current needs of trainees as they progress through their course. By collecting, encoding, analyzing and translating the instructor's feedback into specifications for a GIDS system, the approach will effectively be that of constructing an 'expert system', which will support novice drivers in a way which is directly comparable with the human expert driving instructor.

4.6 Collateral behaviour (multitasking)

Increasingly, drivers will perform in-vehicle activities that are not part of the driving task as such. Carrying on a telephone conversation while driving is just the first step in the direction of using the car as an 'extended office'. The complete GIDS system will have to offer support under these circumstances as well. That is, the GIDS dialogue controller will have to distribute the messages associated with these activities and prioritize them according to urgency, as it does with messages originating from the driving subtasks proper. The necessary extensions of the architecture that should accomplish this are additions to GIDS which its generic nature allows. However, they have not formed part of the development of the GIDS prototype.

4.7 Conclusion

Of the four basic co-driver functions of enhancement, augmentation, interpretation and support the former two are to be considered as the 'classical' ones, i.e., that can largely be executed within the present traffic system. What we hope to have demonstrated in the preceding sections is that the important subtasks that can be distinguished within the driving task – navigation, collision avoidance and control – lend themselves to being subjected to the more advanced functions of interpretation and support, and that these functions can be tuned to the characteristics of individual drivers. This then is what GIDS promises to achieve.

References

Alm, H., Nilsson, L. Järmark, S., Savelid, J., & Hennings, U. (1991). *The effects of landmark presentation on driver performance and uncertainty in a navigation task - a field study.* Swedish Prometheus S/IT-4. Linköping, Sweden: VTI

McRuer, D.T., Allen, R.W., Weir, D.H., & Klein, R.H. (1977). New results in driver steering control models. *Human Factors, 19*, 381-397.

Posner, M.I. (Ed.). (1989). *Foundations of cognitive science.* Boston, MA: MIT Press.

Rockwell, T.H. (1972). Skills, judgement and information acquisition in driving. In T.W. Forbes (Ed.), *Human factors in highway traffic safety research* (pp. 133-164). New York: Wiley-Interscience.

Rühmann, H. (1981). Schnittstellen in Mensch-Maschine-Systemen [Interfaces in Human-Machine Systems]. In H. Schmidtke (Ed.) *Lehrbuch der Ergonomie.* München: Hanser.

Schiff, W., & Foulke, E. (1982). *Tactual perception: A sourcebook.* Cambridge: Cambridge University Press.

Sheridan T. B., & Ferrell, W. R. (1974). *Man-Machine-Systems: Information, control and decision models of human performance.* Cambridge, MA: MIT Press.

Stillings, N.A., Feinstein, M.H., Garfield, J.L., Rissland, E.L., Rosenbaum, D.A., Weister, S.E., & Baker-Ward, L. (1987). *Cognitive science: An introduction.* Boston, MA: MIT Press.

Streeter, L.A., Vitello, D., & Wonsiewicz, S.A. (1985). How to tell people where to go: Comparing navigational aids. *International Journal of Man-Machine Studies, 22*, 549-562.

Verwey, W.B., & Janssen, W.H. (1988). *Route following and driving performance with in-car guidance systems.* Report IZF C14. Soesterberg, The Netherlands: TNO Institute for Perception.

Zwahlen, H.T., Adams, C.C. Jr., & DeBald, D.P. (1988). Safety aspects of CRT touch panel controls in automobiles. In A.G. Gale, M.H. Freeman, C.M. Haslegrave, P. Smith, & S.P. Taylor (Eds.), *Vision in Vehicles II* (pp. 335 - 344). Amsterdam: Elsevier North-Holland.

Part II
The GIDS system

Chapter 5
Design considerations

John A. Michon, Ep H. Piersma, Alison Smiley, Willem B. Verwey, Eamonn Webster

5.0 Chapter outline

The problem the designers of a GIDS system are facing is the explosive complexity of the task space within which the system is operating. Formally GIDS is dealing with a five-dimensional space: *situation* x *task* x *function* x *driver characteristics* x *I/O mode*. In this chapter the constraints imposed by this space on the actual GIDS design are discussed and it is argued that the choices made to reduce the space to manageable proportions do, in fact, retain the fundamental aspects of the GIDS concept. First the considerations underlying the choice of theoretical and practical constraints are elaborated. The next section, Section 5.2, specifies the actual constraints on the GIDS prototype design. This section reviews the situations, task elements, support functions, interface characteristics, and individual factors that comprise the Small World paradigm. In Section 5.3 the constraining factors on the implementation of the GIDS system are discussed. Section 5.4 deals with the concept of adaptivity as a design principle. Adaptivity constitutes the basic feature of the 'intelligence' of the GIDS architecture. The chapter concludes with a brief section, summarizing the boundary conditions for the design of the GIDS prototype. This chapter paves the way for the topics discussed in Chapters 6, 7, 8, and 9.

5.1 The GIDS design philosophy

The systems approach to design

The success of the GIDS project is dependent upon the use of a systems approach to design, an approach that is multi-disciplinary in nature. Both the human and ma-

69

chine aspects of the system, their interaction, and the environment in which they operate have been considered during this design process. The role of the engineer is to consider the possibilities of the hardware and its limitations, such as time to failure, operating environment, tolerance to load, etc. Such considerations have always been seen as important for good design. However, it is only fairly recently that the importance of considering the characteristics of the human user has been recognized. It is still the case that most engineers and industrial designers receive little or no training in ergonomics, and therefore it is hardly surprising that it is still unusual for a product design team to include an ergonomist. In addition, most consumers are unaware that there is a formal body of knowledge in the area of user needs, and their selection of products is not explicitly based on ergonomic criteria. However, this is rapidly changing, as 'user-friendliness' becomes a product characteristic that both manufacturers and customers are more concerned about.

The systems design approach requires a clear definition of the system's goals prior to its design and an evaluation of the design using criteria based on those goals. These should be derived by considering the needs of the system users. The term 'needs' covers two meanings. The first refers to the functions that the driver needs; for example, support in the detection of rapid closure with an obstacle ahead. This is a function that humans perform poorly, the result being many rear-end accidents. The second sense in which the term 'needs' is used relates to human limitations and abilities, for instance the need to compensate for limited information-processing capacity. Goals derived from such needs, rather than from available technology, should drive the design process. For example, it is technically possible to design a head up display for presenting speed information. Whether or not drivers really need speed information to be displayed in this manner is another question. Technical feasibility by itself is an inadequate design criterion!

The role of ergonomics

The role of the ergonomist is to advise the system designer on the human component in the same way that the engineer advises on the hardware and software components, in terms of tasks (functions) performed and in terms of capacities and limitations. What the ergonomist can offer the designer is by no means a complete set of design specifications, but rather a set of boundary conditions and operational ranges within which the designer can be assured that the physical, perceptual, and cognitive abilities and limitations of human operators will be accommodated. The role of the ergonomist complements that of the engineer and the designer. Both are interested in human behaviour and needs, but from different perspectives.

Research on ergonomics and traffic safety has been performed since automobiles were first mass-produced in the 1920s. A great deal is known about many aspects of driver behaviour; for example, the range of reaction times for drivers to perceive unexpected objects on the road and start braking, the amount of time to process information from signs, the average variability of lane position when drunk and when sober, or the average number of times drivers get lost when driving in unfamiliar areas. Apart from the knowledge that is specific to the driving task, a great deal is also known about human performance in general (e.g., Boff, Kaufman, & Thomas, 1986). The latter knowledge will be important in the design of the new vehicle-driver interfaces anticipated by GIDS systems (see also Chapters 2, 3, and 4).

Every new design that changes human behaviour raises new research questions in ergonomics. In particular, some of the relevant knowledge about human behaviour currently exists only in qualitative form. However, for the prototype GIDS system described in the following chapters, more than qualitative information is required. For example, we know that if drivers occasionally receive incorrect information from an interface they will eventually ignore all information coming from that source, because of lack of trust. In the design of an obstacle avoidance system, therefore, the engineer needs to know approximately how many false alarms about impending collisions there can be before the driver begins to ignore all such warnings. Only then can a reasonable warning threshold be set.

One might summarize this systems design approach by saying that it is the role of the ergonomist to establish the user's actual (physical, perceptual, cognitive) needs and it is the role of the engineer to implement a technical response to these needs, either developing new technology or using state-of-the-art technology. It is then the role of the designer to attend to the aesthetic and perceived needs of the user. It is certainly conceivable to do without the designer or the ergonomist in this process. However, the input of the designer will help ensure that the product is marketable and the input of the ergonomist will help ensure that once the product is in use it satisfies the user's needs.

Modularity of design

The general approach adopted in the GIDS project has been one of modularity. One considerable advantage of a modular architecture is that it can deal with the domain to which it is tuned very fast, very specifically, and very reliably. A disadvantage is that it will lack the flexibility of adaptation that is characteristic of a uniform architecture. An important design question was, therefore, which functions and which parts (or subdomains) of the knowledge base were going to be implemented as modules and which should remain so flexible as to require a uniform architecture.

Behaviourally the navigation component of the driving task, the manoeuvring component, the control component, and also the training aspect have been considered separately. This was partly determined by theoretical considerations: the navigation, manoeuvring, and control components represent essentially different subtasks of the driving task, each with different time characteristics and information processing demands, as well as a different level of cognitive control. The training aspect required consideration of additional issues, and was studied by a separate group. More practically, however, the choice of a modular approach was almost equally strongly indicated: the available technology and the knowledge that we possess about the driving task are such that there is currently little hope of incorporating the whole area of driver support in an integrated system.

This deliberate choice in favour of modularity has not, however, resulted in a piecemeal approach to the various design issues. Ideally modularity allows seamless integration between individual modules, simply because it requires that the interfacing between these modules is governed by a uniform information exchange protocol. One of the basic design questions for GIDS has been the specification of such a uniform protocol. Not only does this protocol accommodate the functions (route guidance, anti-collision, and speed and steering control) that will initially be part of the GIDS prototype, it also provides the standard for later additions to the system.

Given that the several subtasks and functions had to be interconnected and simultaneously monitored by the GIDS system, careful consideration had to be given to the format of knowledge about the world in which the user is operating as well as to the handling of this knowledge. Chapters 6, 7, and 8 will describe the actual problems encountered here and the solutions chosen in our attempts to cope with these issues.

Rapid prototyping in the Small World simulation

Despite the fact that our efforts are restricted to the Small World, currently it appears to be a major task to compile a sufficiently detailed inventory of driver performance data to implement the various event representations, and to analyze the dialogue structure of the GIDS system. Yet this constitutes the necessary knowledge base for any driver support system that deserves the description 'intelligent'.

Part of this knowledge base can be derived from the available literature, another part has been collected in the context of the GIDS project. However, these activities provide only the pieces for what essentially resembles a gigantic jigsaw puzzle. The question is to determine the constraints on these pieces so that they will only form legitimate patterns and never allow an illegitimate one to occur. This problem is similar to that of constructing a grammar for a natural language and it involves a considerable amount of trial and error. The first question, therefore, is how to generate and test a sufficiently large set of event sequences to establish and refine the required event representations and scheduling scenarios.

Rather than making this generate-and-test procedure a failure-prone armchair exercise, or a costly real-world experimental programme, the decision was taken at an early stage of the project to implement a Small World simulation. This makes it possible to identify relevant event sequences and critical manoeuvres, within the constraints of the topography and the dynamics of the Small World. It also makes it possible to test all conceivable event representations, message structures, and dialogue scheduling that arise when driving through this (simulated) Small World. This simulation is described in chapter 9. It has greatly expedited the process of implementing the GIDS knowledge base, which depended in large measure on our ability to identify an appropriate set of constraints on what otherwise would be an infinite set of possible, but mostly dangerous, actions. That is, if there were no reasonable constraints, the automaton might be forced to consider such inane actions as braking to a halt in 2 seconds when driving at a speed of 100 km h^{-1}, or turning the steering wheel 180 degrees while driving downhill on an icy road. By selecting the proper constraints, however, such absurd manoeuvres can be avoided on an a priori basis.

GIDS system task functional requirements

GIDS system characteristics will be defined at this point in terms of very general design requirements. It must be emphasized that in this chapter we discuss the concept of GIDS and not the particular prototype that emerged from the GIDS project. Thus the requirements outlined below represent the ideal and not necessarily the immediately practicable or implementable. In later chapters we will consider these characteristics in terms of the actual implementation of the support task. Once a physical form is chosen, for example, verbal instruction for the navigation system, concrete design specifications can be generated. The design requirements given below are derived from a general knowledge of ergonomics and are by no means intended as an exhaustive list.

An understanding of the driver support requirements together with a detailed set of system specifications have allowed predictions to be made about the way in which the use of the GIDS system should affect driver behaviour. These predictions have served as the criteria against which to evaluate the GIDS system. These aspects of the project receive further explicit attention in Chapter 10.

(a) Long-term activity: planning and navigation
Design requirements for the GIDS navigation task should at least include the following points:

— the design should recognize that different drivers will bring to the task different capacities and different a priori knowledge; the design should allow for different degrees of experience and familiarity just as, for instance, word processing packages cater for novices and experienced users;
— the information offered to drivers should conform to their personal mental models of the situation, or supplement these;

- the information should be presented in a form that is meaningful and quickly understood by the driver. It is particularly important to determine whether or not symbols, abbreviations and verbal material are clearly understood;
- the information should be presented at a rate the driver can cope with;
- the information should be adequate to allow the driver to carry out the navigation task successfully;
- the short-term memory capacity of the driver should not be exceeded;
- the system should attract the driver's attention when there is a critical message;
- the presentation of information should take account of current conditions in the driving task and the state of the driver.

(b) Short-term activity: manoeuvring and control

Design requirements for GIDS to support the manoeuvring and control tasks should at least include the following points:

- the range of drivers' reaction times to unexpected situations must be considered, so that warnings can be given in time to allow even the slowest driver to avoid accidents;
- the threshold for activating support should not be set so low that the driver's workload in checking out false alarms increases to a point where safety is compromised, nor so high that valid threats are missed;
- the system should attract the driver's attention under all traffic conditions;
- if a variety of warning signals is given, the driver should easily be able to distinguish between them;
- the support system should be sensitive to current driving task demands and the state of the driver so that warnings are given earlier than normal if necessary;
- automatic action should be considered where a driver would be incapable of responding in time.

(c) Adaptation: parameter settings and instructional feedback

Design requirements for GIDS to support the adaptive parameter setting and instructional feedback functions should at least include the following points:

- over time the default settings for the support functions should adapt to the requirements and capabilities of the individual driver;
- drivers should be able to set and modify default values for the various support functions;
- individual driver data affecting the behaviour of the system should be portable between GIDS-equipped vehicles;

- care must be taken to prevent drivers from inappropriately altering the system's parameters;
- as long as the driver is performing correctly, no feedback should be given, unless requested by the driver;
- if the driver commits an error and the system gives a warning message, an explanation for this warning should be available upon request;
- if the driver repeatedly makes the same error, or persists in undesirable (ineffective or inefficient) behaviour, the system should actively provide the driver with remedial information.

5.2 Constraints

It should be clear that for a long time it will be impossible to incorporate all aspects of the driving task into a real-world driver support system. Formally GIDS is dealing with a five-dimensional space: *situation* x *task* x *functionality* x *driver characteristics* x *I/O mode*. Even if the number of possible states in each of these dimensions were bounded – which may actually not be the case – then the so-called productivity of this five-dimensional space would still be expected to lead to an infinite number of possible trips. A full-fledged support system should be able to offer support in the course of any conceivable trip. Fortunately it is not necessary to design a complete system incorporating all possible states and combinatory rules in order to demonstrate the principle of intelligent driver support. The GIDS concept is generic but its feasibility can be demonstrated in a prototype of limited functionality, operating in a limited number of circumstances. The question is, what constraints will bring the design problem within the limitations imposed by technological, financial, and organizational possibilities, yet allow us to demonstrate the generic character of the GIDS system.

In order to keep the GIDS project within reasonable proportions, a number of constraints were imposed on each of the following five major aspects under consideration:

- driving environment;
- driving task;
- support functions;
- human-machine interface;
- architecture.

Environment

Driving takes place in a physical environment, the road and traffic system, which is part static, part dynamic. Although the GIDS system should eventually be capable of functioning under a wide range of real-world conditions, research is, for the time being at least, restricted to the minimal environment that provides a sufficiently rich domain for studying driver behaviour.

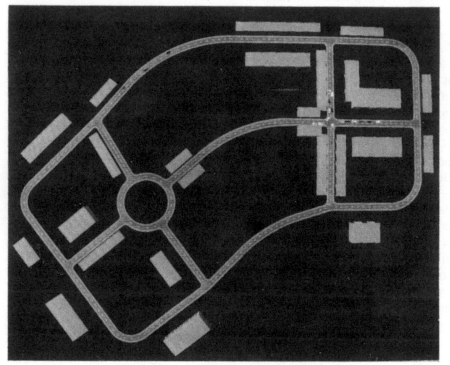

Figure 5.1 The Small World topography

At an early stage of the GIDS project it was decided to restrict the conditions in which the prototype would have to operate to a subset of the normal driving environment. This subset is non-trivial in the sense that, although small, it contains precisely those building bricks from which ordinary road environments tend to be composed, including:

—   roundabout;
—   intersection;
—   straight road sections;

– curved road sections;
– T-junctions.

More than 50 per cent, and perhaps as much as two-thirds of the driving task, consists of coping with just these topographical entities.

This sub-set also comprises what has become known as the Small World, a topographical arrangement entirely composed of elements that belong to the sub-set (Figure 5.1). The environmental characteristics of the Small World also include a number of permanent (two-lane undivided rural road, dry road surface) and semi-permanent features (stationary obstacles, traffic signs), as well as variable elements and conditions (moving obstacles, visibility distance).

Figure 5.2 Perspective view from the road generated by the Small World simulator.

The Small World is small enough to be described exhaustively in formal terms. It can be configured in a computer simulation and on a sophisticated driving simulator (Figure 5.2). In the simulated version of this Small World described in Chapter 9, the characteristics mentioned are the exhaustive set of situational parameters. In real-world conditions other features will, of course, be present and active, but they will not be detected or reacted to by the present GIDS system (see Chapter 10). The Small World can also be realized fairly directly under closed-track conditions. Altogether the Small World provides a set of well-defined environmental conditions for the formal representation of the driving task and for empirical testing of the GIDS prototype.

Driving task

Driving consists of partially ordered sequences of subtasks of the driving task. Various subdivisions have been discussed in Chapter 2. The number of distinct actions involved is of the order of, perhaps, 2000 (McKnight & Adams, 1970; see also Chapter 2). This imposes another, quite limiting, constraint on the range of behaviours that may be incorporated in the repertoire of GIDS in its present form.

Apart from the constraint that the task falls within the normal routines when driving in the situations specified by the Small World, there were three other criteria for inclusion:

— the possibility of generating a sufficiently detailed formal description of the task, the associated support requirements and communication structure (message vocabulary and semantics);
— the possibility of studying the task in sufficient detail and fairly exhaustively;
— the relevance of each subtask for more than one of the support functions.

It should be emphasized that the present set, although meeting these criteria, is not restrictive. Depending on future work, it may have to be modified or extended.

On the basis of these constraints the aim of defining a limited but realistic set of driving tasks was met by including the following activities in the operational GIDS prototype:

— lane following;
— car following;
— overtaking;
— negotiating an intersection;
— negotiating a roundabout;
— merging into a traffic stream;
— exiting from a traffic stream.

Support functions

As was argued in Chapters 2, 3, and 4, the driving task is characterized by three major task performance levels, planning and navigation, manoeuvring, and vehicle control. Again, in order to keep the functional complexity and, in particular, the range of the GIDS prototype within reasonable limits, constraints must be imposed on the number of functions at each level for which support is to be given. In order to attain the goal of demonstrating the generic character of GIDS the present conceptualization of GIDS involves functions at each task performance level: a navigation system for the first and, a collision avoidance system for the second level, plus two control functions, speed and heading control, for the third level.

Apart from these three functional levels that derive directly from the nature of the driving task, there are two more groups of functions to be implemented in the GIDS design: those that support drivers as a function of their personal performance characteristics and those that support collateral activities which are not part of the driving task as such, but which, nevertheless, are carried out quite frequently, sometimes under circumstances that incur a certain hazard. For example, it might be preferable if drivers refrained altogether from carrying on telephone conversations in urban traffic, but such calls will be made anyway, prudent or not, and the GIDS system should therefore be able to support this task. The adaptive functions of GIDS imply, but are not restricted to, adaptation to the specific momentary needs of the driver. The system should also be capable of providing remedial information on the basis of its past experiences with particular drivers and it should also provide tutorial information to inexperienced drivers.

In summary, these considerations have led us to the following choice for support functions to be implemented in the GIDS prototype:

- navigation functions: route guidance;
- anti-collision functions: car following;
- active speed control: accelerator pedal;
- active steering control: steering wheel;
- collateral activities: car telephone and stereo equipment;
- explanatory and tutorial information.

Human-machine interface

The driver will be able to interact with a future GIDS system by means of a variety of displays and controls, in principle using any application which follows the input/output conventions that may be set as a standard GIDS communication protocol. Such a protocol is not yet available.

At the present stage of development, again, there had to be severe constraints on the range of applications actually implemented in the first GIDS prototype. The rationale for a choice of modalities was, again, ergonomic and pragmatic. As stated before, the principal aim of the GIDS project has been to demonstrate the feasibility of the concept of intelligent driver support. In addition GIDS systems should be able to accommodate multiple interfacing systems. The interface components that have been studied in the context of the GIDS project include:

- voice generator;
- speech input recognizer;
- keyboard;
- conventional switches;
- smart card reader (simulated);
- active accelerator pedal;
- active steering wheel.

These components are essentially used in a multifunctional way. Rather than rigidly restricting the output of the GIDS system to specific channels, pertinent messages are presented through selected channels, depending on their momentary availability, the urgency of messages, and the need to optimize the workload of the driver.

Person-related factors

One of the special features of the GIDS system is that it will be able to take into account individual differences between drivers. There are four major categories of such factors:

- age-related;
- experience-related;
- (semi)permanent physical and mental dispositions;
- variable physical and mental dispositions.

In the GIDS project we concentrated on the first two aspects only. Individual drivers may differ greatly as a result of permanent and semi-permanent dispositions, in addition to being subject to relatively quick variations in mood, emotion, or physical well-being. Ultimately GIDS should be able to cope with a variety of such factors, if they can be reliably quantified and if reliable equipment is available to measure their variations. Until such indicators (and equipment) become available, GIDS will not specifically support drivers when they are subject to such variations. Consequently the only individual factors that have been incorporated in the present GIDS system are age and experience. Both can be simply defined and determined and adaptive support can be designed so as to accommodate different groups of drivers: old versus young, or experienced versus novice drivers. Our understanding of the relationships between safe driving and personality or motivation has not yet reached the stage at which it can be formalized to such an extent that these factors can be incorporated into a driver support system such as GIDS.

5.3 Implementation

Bus architecture and peripheral components

The architecture of GIDS, the structure of its actual hardware, is the subject of Chapter 8. Perhaps it would have been possible to build a GIDS system from scratch with all conceivable peripherals, sensors, and applications, specially tailored to the purpose of the GIDS design. The constraints of time, money, and man-power have forced us to rely largely on components that are commercially available. There is, however, a more fundamental reason for using off-the-shelf components and sub-systems for such an approach. Since GIDS represents a generic principle, there should remain a certain independence from specially designed equipment. The strategy of using available state-of-the-art components eminently meets the objective of making the system compatible with non-specialist equipment.

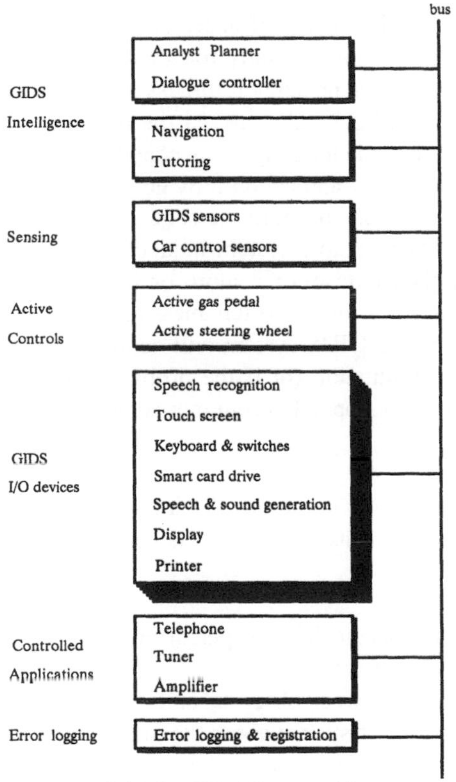

Figure 5.3 The GIDS bus architecture

As will be described in more detail in Chapter 8, a bus architecture has been adopted with the system's dialogue controller as 'server' (Figure 5.3). Rather than completely centralizing all communication on the bus, however, it is efficient to allow certain functions actually to bypass the dialogue control unit temporarily or permanently; for instance, in the case of extremely urgent messages, or when a specific interaction between the driver and a particular support function is required.

Analyst/Planner

The most crucial aspect of GIDS is the software that will enable the system to perform intelligently or adaptively. The general structure of the intelligence of the GIDS system will be described in detail in Chapter 6. GIDS knows about the world, or more specifically, GIDS has a representation of every possible Small World situation and of every type of event that can take place in that world. Each representation consists of a vector of numerical and binary elements (truth values) that unambiguously identifies each event. These elements represent the states of all variables that GIDS will sample through its sensors and applications. Associated with each vector element is a range of acceptable values. As long as drivers' behaviour remains within this range jointly for all variables, there will be no output and they will therefore receive no support, unless they actively *request* information. In that case the system will generally answer the request. As soon as one or more observed variables fall outside the range of acceptable values, GIDS will recognize the pattern of this discrepancy and act on it by generating a message. Every critical situation is associated with one or more messages. The term 'message' is used here in a generic way. It may refer to a system-induced change in accelerator pedal response, a beep, a spoken phrase, or a more or less complicated visual signal: a blinking light, a pictogram, or a written message. Once the message has been passed on to the driver, the GIDS system will continue operating: if the driver recovers from the critical situation (or if the problem goes away as, fortunately, it often does) the system will stop delivering messages until the next critical situation arises.

The GIDS Analyst/Planner combines the inputs from the sensors and applications with the driving task data it already has, in order to perform the following functions:

— recognize what actions the driver is performing;
— anticipate future driver actions;
— detect discrepancies between observed and normal (safe) actions;
— inform other parts of the system about the prevalence of such discrepancies;
— provide the scheduler contained in the dialogue controller with appropriate workload indices.

The Analyst/Planner has a database describing the Small World tasks. These descriptions include the information needed to follow the driver's performance of these tasks. Anticipations of future driver actions are predicted on the basis of previously observed actions in combination with current information about the environment. For this purpose the Analyst/Planner contains rules to detect deviant actions. When such actions are detected the Analyst/Planner informs the other parts of the GIDS system, which then generate relevant information for presentation to the driver at the appropriate time. The Analyst/Planner also provides the Dialogue Controller with workload indices that are computed from the anticipated actions.

Dialogue management

When a message has been generated, the problem is to exchange it with the driver by means of the Dialogue Controller. The first problem to be dealt with is the possibility that several critical events will occur at the same time, or so close in time that a priority conflict arises. The second, related, problem pertains to the driver workload created by the situation and by the message structure. Finally there is the problem of allocation of messages to the various interfaces, in such a way that the driver is not confused or overloaded simply by communication with the system.

On closer inspection the general issue appears to be one of scheduling. Given an event or event sequence, the performance of the driver will depend on when and how to process the relevant information: for example, how long it takes to read a road sign and how long to notice, say, that a child is standing at the kerb waiting to cross. How long does it take to reach a decision to change gear and how long to activate a brake? What is the range of these values in young and elderly drivers, in novice drivers, handicapped drivers...? Many of these basic performance data are known. What has never been studied, however, is how these elementary performance operations fit together into integrated manoeuvres such as merging into a traffic stream or during an overtaking manoeuvre.

In GIDS this problem has been tackled by means of a strategy proposed by Card, Moran, and Newell (1983), based on the concept of the Model Human Processor. Although this approach is of recent origin, it is an immediate descendant of the once popular time-and-motion method of work study. It has been applied successfully to a variety of complex human-machine interactions, including typewriting and various computer operating tasks (John, 1989). It appears to be equally successful for the study of dialogue control in GIDS (Piersma, 1990).

The method as applied consists of determining the time and resource allocation for each separate action unit – turning right, changing lane, braking to a halt at a traffic light, or, for that matter, dialling a number on the car phone – in a particular manoeuvre. A similar analysis is made for each message that may be passed to the driver and for every message that the driver may pass to the GIDS system. These allocation patterns are stored and eventually retrieved together with the various situation and event representations. The function of the dialogue controller is to schedule the actions and messages of the GIDS system in such a way that for every conceivable event or combination of events a nonconflicting interfacing with the driver is maintained. This type of scheduling resembles the application of critical path analysis to industrial management. An interesting but taxing complication is that the GIDS dialogue control requires scheduling for uncertain events in real time, meaning that there is no room for 'what if' strategies for finding optimal schedules. A simple but effective partial solution for this problem has been adopted: events and messages are also assigned a priority value which allows an urgent, potentially life-saving message to overrule schedules of lower priority.

5.4 Adaptivity in perspective

Undoubtedly, much research is required to determine which information in the environment and which driver characteristics correlate most with the workload indices. In the present GIDS version the driver model consists solely of simple, static look-up tables indicating average workload values for groups of drivers in different situations. However, in more sophisticated GIDS versions, production systems or neural networks may contain the driver model. Such models will learn driver idiosyncrasies and adapt to gradual changes in driver behaviour.

One important characteristic of the proposed GIDS system is that it is based on separate modules. Therefore, new applications or subsystems may be connected to the core GIDS system (i.e., the Analyst/Planner, the Dialogue Controller, and several I/O devices) as long as they comply with the GIDS communication protocol. Only when the system can be supplemented quite easily with new applications, will it be feasible to further increase complexity, either during development or after introduction of the system.

Estimating workload

Adaptivity to more or less permanent driver characteristics is incorporated in GIDS by categorization of drivers according to relatively easily detectable features. For the present purpose, these features are age, and driving experience. Future research should examine the precise effects of these variables on driving capabilities and workload and determine whether additional variables should be incorporated in the adaptivity function.

In GIDS adaptivity to dynamic factors is realized by real-time estimation of workload for the driver caused by the present and the imminent driving situation, and by the present and imminent interactions between the driver and the car interface. Thus, dynamic adaptivity requires recognition of the current and forthcoming driving situation, and current and forthcoming information intake and actions of the driver. With the driver model this information can be used to estimate workload for the current driver and to take action when the driver is in danger of becoming overloaded. Such action includes message postponement or the anticipation of messages. Changing the message to a simpler format or code will be incorporated sparingly in the GIDS system, since this carries the risk that workload might be increased due to reduced familiarity.

Because workload is known to be multi-dimensional (Verwey, 1990), estimates of it are expressed in terms of perceptual, cognitive, and effector load on the driver. In order to keep workload under control, ongoing interaction between the driver and a specific GIDS application should terminate before interaction with another GIDS application starts. For example, while the driver is tuning the stereo, the message that the vehicle carries fuel for only another 40 miles should be postponed. However, interactions must be interruptible by messages regarding traffic safety. Therefore, a six-point scale of message priority has been proposed varying from "unimportant for driving safety" to "potentially life threatening" which is used by GIDS to decide whether an ongoing interaction may be interrupted or not.

Admittedly, the presently proposed workload estimator takes no account of the versatility of human behaviour and all the external and internal variables that influence it. Effects of fatigue, lack of sleep, medicines, and drugs are for the present assumed not to affect driver behaviour, which is clearly incorrect. Also, the system is not able to recognize all driver actions. For example, when the driver is looking for a book store or talking to a passenger, workload resulting from these actions is not incorporated in the estimate. These examples serve to emphasize that the driver must remain in overall control and able to switch the system off in the case of perceived overload. This may also serve to increase system acceptance.

Task descriptions

An essential design consideration concerning the 'soft' architecture described in Chapter 6 is that it must operate under real-time conditions. How should GIDS internally represent the driving task, the actual performance of the driver, and the dialogue structure, in order to support real-time action? This depends very much on the format and the properties of the task representation in GIDS. Eventually it should be possible to represent the driving (sub)tasks supported by GIDS as sets of rules in a rule-based system. This would provide maximum behavioural flexibility. For the short term, however, this is a feasible perspective only for some of the navigation and route guidance tasks, and for the tutorial and collateral activity support provided by GIDS. In a variety of manoeuvring tasks, and most evidently in the control tasks with short time constants, existing rule-based intelligent architectures are too slow by several orders of magnitude to offer real-time support, even if it were possible to implement GIDS in really fast computing hardware.

The solution adopted for the present system will be explained in detail later, in Chapter 6. In this approach tasks are described in terms of the subtasks and actions of which they consist, linked together into action networks. These networks represent the relations between actions, the individual nodes in the network corresponding to states in the real world.

5.5 Conclusion

It is not realistic to design a GIDS system in the expectation that it will behave integrally under normal traffic conditions. In this chapter we have reviewed the constraints imposed on the GIDS prototype design. The task space has been reduced to manageable proportions by postulating a well-defined Small World. Although it contains only a few distinct elements, the Small World allows a variety of very mundane navigation, manoeuvring, and handling tasks, so mundane in fact that together they may represent, perhaps, more than half of the average driver's behaviour. In addition the Small World layout can be realized with little difficulty in the real world (Chapter 10) and in simulated conditions (Chapter 9). The adopted constraints can be altered: the GIDS structure is extendible, depending only on the possibility of formalizing a situation, a task, a function, or a driver characteristic. The GIDS prototype to be described in the following chapters therefore retains the principal characteristics of a generic intelligent co-driver support. As such it represents a major step on the way to real time adaptive RTI systems.

References

Boff, K.R., Kaufman, L., & Thomas, J.P. (Eds.). (1986). *Handbook of perception and human performance* (2 Vols.). New York: Wiley.

Card, S.K., Moran, T.P., & Newell, A. (1983). *The psychology of human-computer interaction*. Hillsdale, NJ: Erlbaum.

John, B.E. (1989). Extensions of GOMS analyses to expert performance requiring perception of dynamic visual and auditory information. In *Proceedings of the Conference on CHI,* Seattle, 1-5 April 1990. New York: Association of Computing Machinery.

McKnight, A.J., & Adams, B.B. (1970). *Driver education task analysis.* Final report, Contract No FH 11-7336 (2 Vols.). Alexandria, VA: Human Resources Research Organization.

Piersma, E.H. (1990). *The first GIDS prototype: Module definitions and communication specifications.* Deliverable report DRIVE V1041/GIDS-DIA1a. Haren, The Netherlands: Traffic Research Centre, University of Groningen.

Verwey, W.B. (1990). *Adaptable driver-car interfacing and mental workload: A review of the literature.* Deliverable report DRIVE V1041/GIDS-DIA1. Haren, The Netherlands: Traffic Research Centre, University of Groningen.

Chapter 6
GIDS intelligence

Henry B. McLoughlin, John A. Michon, Wim van Winsum, Eamonn Webster

6.0 Chapter outline

In this chapter we consider the intelligent components of the GIDS system. We will begin by arguing for the necessity of having such an intelligent component (Section 6.2). We show how it fits into the overall architecture of GIDS. In the subsequent Sections 6.3 to 6.6 we describe the approach to building this component, showing what knowledge is required and how it was acquired. We provide examples of the knowledge and how it is represented and used. Finally, in Section 6.7 we speak of the generic nature of the GIDS system and in Section 6.8 we examine future developments.

6.1 Introduction

To speak of a vehicle as having knowledge and being capable of reasoning about its knowledge may at first seem quite a strange idea. A number of questions come immediately to mind: What sort of knowledge might it have? How would it represent this knowledge? What sort of reasoning might it perform? Indeed, more serious philosophical questions also arise. Can we really say that an artefact can have intelligence, or is this just a flight of anthropomorphic fantasy? In this chapter, we acknowledge the existence of such philosophical ramifications but we deliberately choose to ignore them. The interested reader is referred to Kelly for a full coverage of the issues (Kelly, 1990). Instead, we will simply retain the currently accepted vocabulary of Artificial Intelligence and speak of Knowledge and Intelligence and concentrate on trying to answer the first set of questions which we have posed.

6.2 The need for intelligence in a vehicle

Intelligent driver support is required in view of the increasing information load on vehicle operators. Clearly, the number of information-providing devices in a vehicle is increasing and there is no reason to believe that this situation will change. Whilst the motivation for providing such devices is laudable; namely to increase the amount and indeed quality of information available to operators, in order that they may be able to make more informed decisions, the end result may be quite the opposite if, by using a number of devices, we provide too much information. This will increase the information processing load on the operator and thus reduce the amount of processing resource available for dealing with the normal driving information provided by the environment. At best this will result in the driver choosing to ignore the various devices; at worst it will overload the driver and result in compromising safety. Clearly this is undesirable and it defeats the purpose of introducing such devices in the first place.

If we consider such devices as experts in limited domains we can hardly expect them to utilize knowledge that lies outside their domains. Yet, there are times when such additional knowledge would be crucial. Consider, for example, a vehicle equipped with both a headway detector and a course-control device. The former acts to warn the driver if he is approaching too close to the vehicle in front. The latter detects when the vehicle moves out of its current lane and warns the driver. If we consider the behaviour of both of these devices when the driver is performing an overtaking manoeuvre, we may find that both of them are issuing warnings as the vehicle approaches the one in front and moves out of lane. Clearly, this would be an undesirable situation.

Ideally, such devices should be tolerated in the vehicle because they can provide valuable information. But, to avoid overloading the driver with too much information, or contradictory, or useless information, what is needed is a component that has an overall view of the driving task and that can use this view to decide whether it is appropriate to pass on a message from a particular device to the driver.

Let us consider yet another scenario. In the course of a trip a driver will encounter a number of different driving situations. Some may be performed quite effortlessly, whilst others will require considerable care and concentration. This variation in workload is something which the information-providing devices will be unaware of. This can lead to the same information being sent to the driver in different situations having different effects at different times. In the case where the current workload is low the information may be beneficial; where the workload is high it may be ignored or, failing that, it may actually prove to be dangerous.

Once again it seems clear that what would be required would be a component which had an overall view of the current driving situation together with an estimate

of the current workload, which would decide on the appropriateness of communicating information to the driver. We should also bear in mind here that such a component would ideally be able to distinguish between the levels of expertise of the
driver in order to function effectively. In a given driving situation the workload experienced by a novice driver may be considerably higher than that of an experienced driver.

Our conclusion from this section is that there will be a continued proliferation of
information-providing devices in vehicles. These devices by themselves may indeed try to provide helpful information to the driver. But, in combination their independent support runs the risk of compromising the safety of the driver. Therefore, what would seem to be needed are components which can take an overall
view of both the current driving situation and the driver's behaviour and use this
'insight' to decide on the apropriateness of advising the driver of the output of the
individual devices.

Clearly, such a component will have to display a certain degree of intelligence
and must have considerable knowledge of the driving task, of the current situation,
and indeed of the individual driver. In the next section we will explore the feasibility of providing such components.

6.3 Modelling the driver

In this section we provide a brief historical review of attempts to model the behaviour of the driver. This work clearly has its roots in traffic science and psychology
but similar developments have occurred in the sub-field of artificial intelligence
(AI) known as User Modelling. The latter draws heavily on work on person perception and sociology and has had a profound influence on our thinking. We will trace
this work to arrive at what was the first conceptual architecture of the GIDS system.

The first serious driver models date from the mid 1960s (for critical reviews see
Michon 1985, 1989). These were the dynamic servo-control models that allow fairly precise predictions of lane keeping and car following performance under ideal
conditions. More or less simultaneously there were early attempts at information-
processing models, although none of these was adaptive.

At that time some adaptiveness was displayed, however, by the precognitive
loop models. A precognitive loop is neither more nor less than a switch. It is sensitive to a set of specific external conditions, for instance a change from dry to icy
road conditions. It is also tuned to performance errors becoming larger than a
preset acceptance threshold. Whenever one of the critical set of conditions is satisfied, the switch will operate, thereby resetting the parameters of the model so as to
achieve an optimal performance of the model under the new conditions. Although

the precognitive loop dates back to the early days of dynamic control modelling, we can equally well regard it as an early attempt at cognitive modelling:

>   if condition X prevails
>   then initiate parameter settings (Y1,Y2,...Yn)

This 'if-then' format illustrates that the precognitive loop is, in a trivial sense, a rule-based representation.

In general, neither the dynamic control models nor the early information processing models – nor, for that matter, the simple precognitive loop system – satisfied the requirements of a computational theory (see Boden, 1989) of driver behaviour. Only as late as 1984 can one find a serious proposal for such a computational approach to the problem of formally describing the driver task (Michon, 1985). Even then it took another two years before the initiation of a research programme aimed at a genuine cognitive approach to driver behaviour (Michon, 1987). This research programme ultimately rested on two premises: (a) the availability of a sufficiently detailed analysis of the driving tasks, namely the one developed by McKnight and Adams (1970; see also Chapters 2 and 3 in this volume). The work of these authors has long been undervalued despite the fact that it gives a truly indispensable overview of the driving task in 1700 elementary subtasks; (b) the availability of sufficiently well-tested "intelligent architectures" based upon the production system concept as embodied in, among others, ACT*, Soar and AB-STRIPS (Akyürek, 1992), and, in addition, the work on planning due to authors such as Tate or Vere (see Allen, Handler, & Tate, 1990).

In parallel with this previous work in traffic science, a sizeable effort has been devoted by researchers in AI since the early 1970s to modelling human social behaviour. The early efforts of Schank and Abelson (1977) inspired a considerable amount of research which mainly focused on elaborating the model of behaviour based upon scripts and plans. Unfortunately the majority of this work centred around using these concepts to understand natural language narrative. Little attention was paid to actually planning in a dynamic environment, or generating dialogue appropriate to a particular person. No explicit attempt was made to model the characteristics of particular individuals. Although Schank and his team did toy with the idea of developing user models in the late 1970s, nothing substantive came from the work.

The first serious attempts at modelling users came from the Toronto group of Perrault, Cohen and Allen (Cohen & Perrault, 1979; Perrault & Allen, 1980). Their aim was to develop a computational formalization of Searle's reclassification of illocutionary acts based upon the pioneering work of Austin. In this work they developed explicit models of the beliefs of other people; these they termed user models. The models were quite complex and allowed for deep nesting within models. This allowed them to reason about the user's beliefs about the system's beliefs about the user. Unfortunately, the aims of this research were to study quite specific linguistic phonemena and, as a consequence, although the structure of the user model was quite complex, its content tended to be rather small and offered a very limited characterization of the user.

A significant development was the Grundy system (Rich, 1980). This offered a wider characterization of a user in terms of a set of attribute value pairs. The values ranged from 1 to 1000 with high values indicating that the user possessed that particular attribute to a high degree. The set of attributes was extremely large. The system was used to select appropriate reading material for the particular user and was able to distinguish between levels of educational background, typical interests, etc. Unfortunately, the user model relied on the system being able to interpret its contents and so the real intelligence of the system resided in the program and the model simply acted as data.

The majority of user modelling work in the 1980s tended to concentrate on the use of such models to explain pragmatic phenomena and is of little practical value in our modelling of the driver. However, in 1987 McLoughlin proposed a new type of model called a Persona, which modelled the user explicitly in terms of the typical tasks which they would be expected to perform normally in particular situations (McLoughlin, 1987). This new model was heavily influenced by the work of the sociologist Goffman (1959). It was used to monitor different types of users interacting with a subset of the VAX/VMS operating system. It allowed the system to detect deviant behaviour and provide advice appropriate to the level of expertise (McDermott, McLoughlin, & Toland, 1989). The model could be adapted by the system to reflect the user becoming more familiar with the system and therefore moving to a higher level of expertise. The success of this work suggested to us that if we could model drivers' tasks in a similar fashion then we would be able to correctly monitor their actions and advise if deviant behaviour occurred. In conclusion, given the results of this previous research, we were able at the start of the GIDS project to propose an initial conceptual architecture which incorporated a model of normative driving and we felt confident that in the course of the work we could elaborate on this model and its use. The figure from Smiley and Michon (1989) illustrates our initial position (Figure 6.1). In this figure, the reference driver model indicates the expected or appropriate driver behaviour given the current environment. The actions of the driver, together with the current environment, provide a model of the actual behaviour of the driver. The Analyst compares these two models and if it detects discrepancies between expected and actual behaviour, it can instigate remedial actions depending on how serious these discrepancies are.

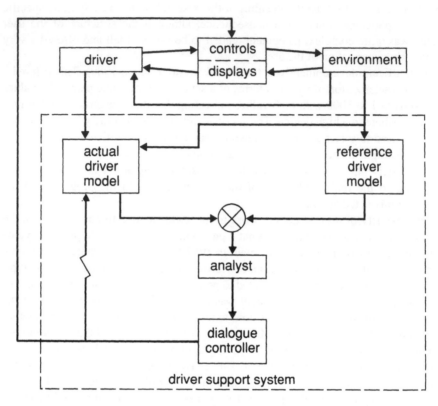

Figure 6.1 GIDS system basic components (after Smiley & Michon, 1989)

6.4 Task analysis

The development of the GIDS system requires a model of the car driver. This model should describe what action is to be expected from drivers in what situation. The system needs to know their expected actions in order to assist the drivers if they fail to act appropriately. The system also needs to recognize the situation drivers are in, in order to evaluate whether a certain action is required of them. So, to enhance the safety of drivers we expect the GIDS system to support them in situations where they fail to perform the appropriate action.

In the early conceptions of the GIDS system a situation was simply a task such as, for instance, car following, negotiating an intersection, lane changing, or overtaking. The idea was that the situation drivers were in, that is, the task they were performing, could be derived in part from their actions, because a task can be regarded as a chain of actions with a certain temporal ordering. The exact nature of this temporal ordering had to be extracted from the literature, in particular from task analyses, such as the normative task analysis of McKnight and Adams (1970).

If, from previous actions, the system knew what task drivers were performing, it could make predictions of their actions. If they failed to perform the predicted action, the system had to intervene by warning them. So, in this sense, the model of the driver consists of an ordered list of actions for every task. Of course, in this conception of a driver model, the actions of drivers are not enough to resolve uncertainties concerning the task they are performing. The system also needs to know something of the environment around the driver, for instance if there is an intersection ahead, a car in front, etc. It was expected that this information on the environment, together with the list of required actions, could resolve uncertainties with respect to the task in which the driver is engaged.

The second important question centres around actions. In the literature on task analyses an action is contingent upon the task which consists of a chain of actions. An action then depends on a previous action, but also on a circumstance occurring during the execution of the task. In McKnight and Adams (1970), a very simple chain of actions during the task of negotiating an intersection might be:

action : look to the right
   if (circumstance : no traffic from right)
   then action : drive on
   else if (circumstance : traffic from right and nearby)
   then action : stop

A detailed analysis of the manoeuvre of intersection negotiation has been carried out at the Traffic Research Centre (Aasman, 1988; Aasman & Michon, 1991, 1992; De Velde Harsenhorst & Lourens, 1987).

The Small World

First generation GIDS systems are required to offer driver assistance in a subset of real-world situations. This subset of situations has become known as the Small World (see also Chapters 1, 5 and 9). Although the Small World contains the majority of relevant real-world situations, it has some restrictions that reduce its complexity. It consists of straight road sections, curves with differing radii, T-junctions, crossroads and roundabouts. The roads are dual lane and allow two-way traffic. Cars are the only type of traffic participants and they behave according to standard rules. If the driver is on a straight road section, the manoeuvres are stopping, moving off, avoiding an obstacle, and overtaking. Manoeuvres in a curve are decelerating and avoiding an obstacle. On a T-junction, at crossroads and a roundabout the manoeuvres are decelerating, using indicators, turning and yielding for traffic, and changing lanes on a roundabout. This implies that the required tasks to be handled by the GIDS system are car following (and general obstacle avoidance), lane changing, overtaking and negotiating intersections (see Chapter 5).

Some properties of the task analysis of McKnight and Adams

McKnight and Adams (1970) have described these tasks in a sequential manner. Overtaking (task 32) for example is a complex task which also contains obstacle avoidance (car following, avoiding a collision with the vehicle behind and with an approaching vehicle) and lane changing. Negotiating intersections (task 41) consists of avoiding a collision with the vehicle behind, changing lane, signalling, avoiding a collision with vehicles approaching the intersection and, if there is a lead vehicle, avoiding a collision with it. So the task organization is hierarchical.

The task analysis of McKnight and Adams is particularly useful in its description of the steps to be taken during a certain manoeuvre, that is, its temporal ordering. A task consists of a series of subtasks made up of chains with the following general structure :

(1)   check whether situation A is true
(2)   if situation A is true
         perform action 1
         else perform action 2.

Although the description of the situations is very detailed and thorough, it is not complete and not specific enough for our purpose in one sense, and too redundant for our purpose in another sense. This will be explained using a description of a task according to McKnight and Adams (Task 34 lane changing) as specified in Table 6.1.

Evidently this description is not complete. Condition 34-11 checks whether a lane change is allowed or safe. It states that a lane change is legally allowed if there are no signs or markings prohibiting a lane change. Condition 34-12 states that a lane change is safe if there is no rear vehicle in the other lane, or there is no rear vehicle in the same lane which is about to enter the other lane, that might be hindered. It does not, however, say anything about the possibility that there might be a lead vehicle in the other lane, that is close to the car, and decelerating. In that case it would probably be very unwise to change lanes. It is also not recommended to change lanes if there is an approaching vehicle in the other lane which is closing in very fast on the car. Task 34 does not mention this, although it is mentioned in task 32 (passing/overtaking). So, the description of lane changing does not completely specify objects that may endanger lane changing. Apart from being insufficiently comprehensive, the description is not specific enough for our purposes. The subtasks 32-1211 and 34-1212 do not specify when the driver is allowed to change lane. It only suggests that it has something to do with the distance and relative speed of the rear vehicle. The same criticism applies to 34-1213. If a rear vehicle signals to change lanes, it does not automatically imply that the car is not allowed to change lanes. There are also actions which appear to be unconditional. Task 32-22, for instance, does not specify the circumstances that require the driver to accelerate.

Table 6.1 - Description of Task 34 (lane changing) according to McKnight and Adams (1970)

34-1	- DECIDE TO CHANGE LANES
34-11	- Determines whether lane change is legally permissible
34-111	- Looks for regulatory signs prohibiting lane change well in advance of manoeuvre
34-112	- Observes pavement markings
34-12	- Looks for rear-approaching traffic in new lane
34-121	- Checks rearview mirror to observe:
34-1211	- Vehicles passing in new lane
34-1212	- Following vehicles closing fast from rear in new lane
34-1213	- Following vehicle about to enter new lane
34-122	- Looks out window to check blind spot, moving head enough to see around blind spot
34-1221	- Varies speed of car very lightly to help bring into view any vehicle travelling in the blind spot at exactly the same speed as the car.
34-123	- On multi-lane roads, looks for vehicles about to enter new lane from the far adjacent lane
34-2	- PREPARES TO CHANGE LANES
34-21	- Signals intention to change lane by activating directional signal and/or employing appropriate hand signal
34-22	- Adjusts car speed
34-221	- Accelerates if possible or maintains speed
34-3	- CHANGES LANE
34-31	- If possible, waits a few seconds after signal before beginning lane change
34-32	- Turns wheel and enters new lane
34-4	- COMPLETES LANE CHANGE
34-41	- Positions car in the centre of new lane
34-42	- Cancels directional signal
34-43	- Adjusts speed to traffic flow in new lane

In another sense the task descriptions of McKnight and Adams are redundant. In the passing task (32) a number of situations determining whether overtaking is allowed are mentioned that are actually equivalent to a number of situations mentioned in the lane changing task (34). There are also descriptions of actions in the passing task which are mentioned in the lane changing task as well, for instance, "signals lane change well in advance of manoeuvre." This is because overtaking is not allowed if changing lanes is not allowed. There are also a number of things mentioned in the task description of overtaking which should have been mentioned in the description of changing lanes; for instance, Task 32-24 maintains proper following distance prior to lane change. However, this subtask is referred to as also belonging to task 31, car following.

The conclusion is that in the description of McKnight and Adams a number of tasks are performed at about the same time, making the descriptions complex and redundant. This is perfectly understandable, because the task analyses of McKnight and Adams have a different purpose from ours. A number of the shortcomings of this task analysis for our purpose simply result from the fact that it is not generic.

The task organization proposed by McKnight and Adams is intuitively very sound and complies with the sequences of actions that are required during car driving. We will argue, however, that it is not always the right sequencing of actions that makes driving safe, but the adequate selection of lateral position and speed. The sequence of actions only makes a difference for other traffic participants, meaning that warning other traffic participants might be appropriate before one changes one's behaviour. Signalling is in fact the only action that has a before-after relation with other actions. Apart from this, the temporal ordering in the task analysis of McKnight and Adams mainly consists of the following basic ordering

(1)   check whether action A is allowed
(2)   choose appropriate parameters for action A
(3)   perform action A

where 1 and 2 are not really actions. Item 1 is a condition, item 2 specifies how the action is to be performed, and only item 3 is the execution of the action. Whilst all three of the above are tasks performed by the driver, the third is the only one which can be detected by the GIDS system. The system could perform the first two tasks but could never monitor the driver performing them. Clearly, we needed to re-examine the McKnight and Adams analysis of Small World tasks so as to isolate detectable driver actions.

The nature of actions

The re-examination of the work of McKnight and Adams (Webster, Toland, & McLoughlin, 1990) yielded the following categorization of detectable driver tasks:

(a) Signalling actions
   – turn on directional indicator;
   – turn off directional indicator;
   – flash headlights;
   – sound horn.

(b) Steering actions
   – swerves sharply to the right to reduce impact angle;

- position car to centre of new lane;
- centre car in new lane;
- enter new lane;
- return to driving lane;
- enter roundabout on right hand side;
- enter inside lane;
- exit roundabout;
- continue around roundabout again;
- begin left turn before centre of intersection;
- turn into far left lane;
- turn into near right lane;
- follow vehicle;
- enter transit lane closest to kerb.

(c) Acceleration actions
- accelerate;
- accelerate quickly to get out of the way;
- accelerate (do not exceed speed limit unless necessary);
- adjust speed;
- slow down;
- maintain speed;
- move through blind spot quickly.

(d) Waiting actions
- wait a few seconds;
- wait until vehicle begins turn.

(e) Yield actions
- yield to traffic on roundabout;
- yield to vehicle in intersections;
- yield to approaching traffic;
- yield to approaching vehicle.

(f) Stop actions
- stop/yield to approaching vehicle;
- stop in advance of other vehicle.

(g) Calculation actions
- estimate speed and distance of approaching vehicle;
- estimate passing distance;
- estimate required acceleration for overtaking;
- check traffic from left.

The GIDS system will only be able to monitor safe performance of the driver if the system can and does calculate whether an action is within safe limits. This implies that calculation actions must be carried out by the GIDS system. Thus for GIDS purposes, calculation actions are not driver actions but GIDS actions. Accelerative actions, yield actions, and stop actions have in common that they refer to a required velocity. Accelerative actions result in a new velocity. Stop actions result in a required velocity of zero, and yield actions result in a velocity which is lower than or equal to the current velocity. Wait actions refer to not changing the current velocity. Together these classes of actions can be regarded as speed actions.

This leaves us essentially with three classes of actions: signalling actions, steering actions, and speed actions. Of these actions, only signalling actions have a clear temporal relation with steering actions. If the lateral position is to be changed there must be a signalling action before the lateral position is changed in order to warn other drivers of the possibility that their path will be crossed by the car. There is no forced temporal relation between signalling actions and speed actions, except for the brake lights signalling if the deceleration is greater than a certain value.

A signalling action is a discrete action occurring before a steering action of a certain magnitude is executed. If no change in lateral position is intended and if the signal is on, the signal must be switched off.

A steering action results as the difference between the current lateral position and the required lateral position or as the difference between the current wheel angle and the required wheel angle (to keep the lateral position unchanged if we are negotiating a curve or to bring the car onto another path if we are negotiating an intersection).

A speed action results as the difference between the current speed and the required speed.

The actions have parameters, consisting of the difference between the actual state and the required state.

The current velocity, wheel angle, and lateral position are given. This leaves us with the problem of specifying the required velocity, wheel angle and lateral position.

Despite all the high-level categorization of tasks, the essential task being performed is the control of the velocity, wheel angle, and lateral position of the car. Any attempt to support the driver's execution of higher level tasks that does not concern itself with these low-level tasks is bound to fail. The choice of appropriate values for these parameters varies with the circumstances. The higher level tasks are descriptions of different circumstances·and provide constraints on appropriate choices.

6.5 Situation analysis

In the previous section we described how we arrived at a definition of the driving tasks which can take place within the confines of the Small World. In this section we describe the driving situations in which these tasks actually take place. We provide here an exhaustive list of the situations which we are currently catering for in the Small World and then give one example of how these situations are represented as rules. We justify giving the exhaustive list because it shows how rich a set of situations can be dealt with even within the confines of the Small World. On the other hand we restrict ourselves to providing just one example of a situation encoded in the rule base and refer the interested reader to Van Winsum (1991) for a fuller description of the rules. In the descriptions that follow the supported automobile will be referred to as the *car*; other traffic will be referred to as *vehicle(s)*.

A situation consists of a certain constellation of:

- positions of other objects relative to the car;
- distances of other objects from the car;
- velocities of other objects relative to the velocity of the car;
- additional properties of other objects;
- properties of the car.

The following (continental) situations may require the system to intervene:

Car following
(1a)   The car is too close to a vehicle in front and is in the same lane.

Rear vehicle
(2a)   The rear vehicle is close to the car which is decelerating too hard.

Approaching an intersection regulated by a traffic light
(3a)   The car is too close to an intersection and traffic light is red.
(3b)   The car is too close to an intersection and traffic light is amber.

Negotiating a curve
(4a)   The car is too close to a curve and velocity is too high for that curve.
(4b)   The car is in a curve and velocity is too high for the curve.

Maximum speed
(5a)   The speed of the car is higher than the speed limit.

Overtaking traffic

- (6a)   The rear vehicle is close to the car and has its left indicator on and the car is accelerating.
- (6b)   The rear vehicle is close to the car and is in the lane left of the car which is accelerating.
- (6c)   The car is too close to a vehicle in front and is in the lane to the right of it.

Right of way rules at intersections

- (7a)   The car is close to an intersection and vehicles on the road it is going to cross or going to merge with have right of way (e.g., the car approaches a roundabout or is on the side road of a T-junction) and the first vehicle approaching the intersection from the left branch has its direction indicator off, or is indicating left and that first vehicle from the left is within a certain distance or time range and the car is going too fast to yield right of way.
- (7b)   The car is close to an intersection and is currently not on a right of way road, it intends to turn left at the intersection or go straight ahead (cross the intersection) and the first vehicle from the right is within a certain distance or time range and the car is going too fast to yield right of way.
- (7c)   The car is close to an intersection intending to turn left and the first approaching vehicle from straight ahead is within a certain distance or time range and that vehicle turns right or goes straight ahead and the car is going too fast.
- (7d)   The car is close to an intersection intending to turn left and the first approaching vehicle from straight ahead is within a certain distance or time range and that vehicle turns left.

Conflicting traffic approaching an intersection does not comply with the rules

- (8a)   The car is close to an intersection and the first approaching vehicle from the left is closer to the point where the two will meet than its braking distance and the car is also closer to the point where the two will meet than its braking distance.
- 8b)   The car is close to an intersection and the first approaching vehicle from the right is closer to the point where the two will meet than its braking distance and the car is also closer to the point where the two will meet than its braking distance and the car intends to turn left at the intersection or go straight ahead.
- (8c)   The car is close to an intersection and the first approaching vehicle from ahead is closer to the point where the two will meet than its braking distance and the car is also closer to the point where the two

will meet than its braking distance and the car intends to turn left at the intersection, or the approaching vehicle is indicating left.

Overtaking

(9a) The car is in the right lane behind a vehicle; there is a vehicle approaching from the opposite direction and the time it would take to move into a safe gap in front of the lead vehicle is less than the time within which the car reaches the approaching vehicle and it is possible to overtake if the car accelerates and indicates left.

(9b) The car is in the right lane behind a vehicle and the time for which a safe gap exists in front of the vehicle ahead is less than the time within which the car would reach that gap but it is possible to reach that gap in time if the car accelerates sufficiently and indicates left.

(9c) The car is in the right lane behind a vehicle and the time for which the distance from a safe gap to the next intersection is longer than a safe stopping distance is less than the time within which the car would reach that gap but it is possible to reach that gap in time if the car accelerates sufficiently and indicates left.

(9d) The car is in the right lane behind a vehicle and there is a vehicle approaching from the opposite direction and the time it would take to move into a safe gap in front of the lead vehicle is less than the minimum time within which the car would reach the approaching vehicle and the car indicates left.

(9e) The car is in the right lane behind a vehicle and the time for which a safe gap exists in front of the vehicle in front is less than the minimum time within which the car can reach that gap and it indicates left.

(9f) The car is in the right lane behind a vehicle and the time for which the distance from a safe gap to the next intersection is more than a safe stopping distance is less than the minimum time within which the car can reach that gap and it indicates left.

(9g) The car is in the left lane behind a vehicle in the lane to the right of the car and there is a vehicle approaching from the opposite direction and the time it would take to move into a safe gap in front of the lead vehicle is less than the time within which the car will reach the approaching vehicle but it is possible to overtake if the car accelerates.

(9h) The car is in the left lane behind a vehicle in the lane to the right of the car and the time for which a safe gap exists in front of the lead vehicle is less than the time within which the car would reach that gap but it is possible to reach that gap in time if the car accelerates sufficiently.

(9i) The car is in the left lane behind a vehicle in the lane to the right of the car and the time for which the distance from a safe gap to the next intersection is more than a safe stopping distance is less than the time within which the car would reach than gap but it is possible to reach that gap in time if the car accelerates sufficiently.

(9j)   The car is in the left lane behind a vehicle in the lane to the right of the car and there is a vehicle approaching from the opposite direction and the time it would take to move into a safe gap in front of the lead vehicle is less than the minimum time within which the car will reach the approaching vehicle.

(9k)   The car is in the left lane behind a vehicle in the lane to the right of the car and the time for which a safe gap exists in front of the lead vehicle is less than the minimum time within which the car can reach that gap.

(9l)   The car is in the left lane behind a vehicle in the lane to the right of the car and the time for which the distance from a safe gap to the next intersection is more than a safe stopping distance is less than the minimum time within which the car can reach that gap.

Lane change

(10a)   The distance to the next intersection is not too short; the car will turn left at the next intersection and it is indicating left and is in the right lane and the distance from the rear vehicle in the lane left of the car is too short.

(10b)   The distance to the next intersection is not too short; the car will turn left at the next intersection and it is indicating left and is in the right lane and the distance from the lead vehicle in the lane left of the car is too short.

Obstacle avoidance

(11a)   The car is in the right lane too close to an obstacle ahead and the distance from the rear vehicle in the lane left of the car is too short.

(11b)   The car is in the right lane too close to an obstacle ahead and the distance from the lead vehicle in the lane left of the car is too short.

(11c)   The car is in the right lane too close to an obstacle ahead and the distance from the rear vehicle in the lane left of the car is not too short and the distance from the lead vehicle in the lane left of the car is also not too short.

(11d)   The car is in the left lane too close to an obstacle ahead and the distance from the rear vehicle in the lane to the right of the car is too short.

(11e)   The car is in the left lane too close to an obstacle ahead and the distance from the lead vehicle in the lane to the right of the car is too short.

(11f)   The car is in the left lane too close to an obstacle ahead and the distance from the rear vehicle in the lane to the right of the car is not too short and the distance from the lead vehicle in the lane to the right of the car is also not too short.

Lane keeping

(12a) The time to line crossing is less than x seconds and the line that will be crossed within x seconds marks the right-hand kerb of the road.

(12b) The time to line crossing is less than x seconds and the line that will be crossed within x seconds marks the left-hand kerb of the road.

Moving off

(13a) The car is moving off, and the distance from a rear vehicle in the lane to be joined is too short.

Navigation

(14a) The car is close to an intersection at which it will turn left and it is not indicating left.

(14b) The car is close to an intersection at which it will turn right and it is not indicating right.

(14c) The car is close to an intersection at which it will drive straight on and it is indicating left or right.

6.6 Architecture of the system

The overall architecture of the GIDS system, and the functions provided are described in detail in Chapters 7 and 8 of this book. In this section we outline the components which make use of the knowledge of the driving task and the driving situations described earlier in this chapter. The two main components are the Analyst/Planner and the Dialogue Controller. These are responsible for determining the current driving situation and filtering messages that are to be passed to the driver. Figure 6.2 shows where these components fit into the overall architecture.

Situation descriptions

The situation descriptions listed in the previous section are encoded into a rule base and these rules are used by the Analyst/Planner to perform its tasks. As an illustration we outline here the rules which encode the overtaking task. A basic requirement is that we can determine whether the car is overtaking or not. The overtaking task can be divided into three phases, pulling out, passing and pulling in. The phases are characterized by the spatial and velocity relationships between the car and the vehicle it is overtaking.

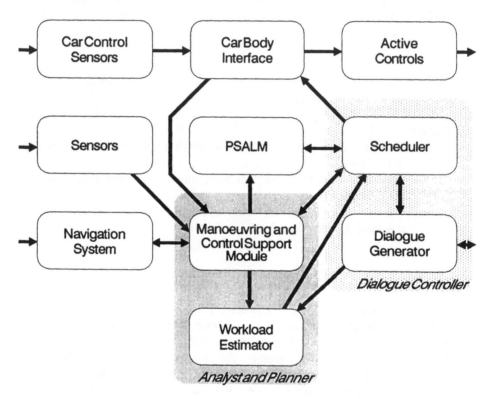

Figure 6.2 The GIDS architecture (for details see Chapter 8)

To determine whether the car is overtaking we must take into account the three phases. Pseudo-code for determining whether the car is overtaking is as follows:

Variables:
    ReactionTime (seconds):
        A measure of the reaction time for the Car
    ReactionDistance (metres):
        A measure of the reaction distance for the Car
    Car (vehicle):
        The car that we are determining whether it is overtaking

RearVehicle (vehicle):
> The vehicle travelling in the same direction as the Car whose centre is behind the Car's centre

LeadVehicle (vehicle):
> The vehicle travelling in the same direction as the Car whose centre is in front of the Car's centre

OvertakeVehicle (vehicle):
> The vehicle that the Car is overtaking, one of (RearVehicle, LeadVehicle, *null*)

MMCO:
> Murphy's Multiplication Constant for Overtaking, i.e., a value that you multiply to get the desired result. The value used was 4. The higher the value, the more cautious the rules become. The value 4 was arrived at through various trials.

Functions:

Lane (vehicle):
> returns the lane the vehicle is in, one of ('Left', 'Right', 'LeftShoulder', 'RightShoulder'), from the viewpoint of the Car

LeftOf (Lane):
> returns the lane to the left of Lane

ChangingLane (vehicle):
> returns lane changing information about vehicle, one of ('ChangingToLeft', 'ChangingToRight', 'NotChanging')

Indicator (vehicle):
> returns the state of a vehicle's indicators, one of ('Left', 'Right', 'None')

Distance (vehicle1, vehicle2):
> returns the longitudinal distance from the front of vehicle1 to the rear of vehicle2 (metres)

Speed (vehicle):
> returns the speed of the vehicle (metres per second).

Rules:

The Rules that determine whether the Car is overtaking do so by returning the vehicle that the Car is overtaking. A *null* value indicates that the Car is not overtaking.

if   (
            LeadVehicle == *null and*
            RearVehicle == *null*
)

```
            {
                    // No lead vehicle or rear vehicle, hence can't be overtaking
                    OvertakeVehicle = null;
            }
else if     (
            LeadVehicle and
            Lane (Car) == 'RightLane' and
            ChangingLane (Car) == 'ChangingToLeft' and
            Speed (LeadVehicle) < Speed (Car) and
            Distance (Car, LeadVehicle) < 2 x MMCO x ReactionDistance
            )
            {
                    // Pulling out into the left lane to overtake lead vehicle
                    OvertakeVehicle = LeadVehicle;
            }
else if     (
            LeadVehicle and
            Lane (Car) == LeftLane and
            ChangingLane (Car) == 'NotChanging' and
            Speed (LeadVehicle) < Speed (Car) and
            Distance (Car, LeadVehicle) < MMCO x ReactionDistance
            )
            {
                    // In left lane passing lead vehicle
                    OvertakeVehicle = LeadVehicle;
            }
else if     (
            RearVehicle and
            Lane (Car) == LeftLane and
            ChangingLane (Car) == 'NotChanging' and
            Speed (RearVehicle) < Speed (Car) and
            Distance (RearVehicle, Car) < MMCO x ReactionDistance
            )
            {
                    // In left lane passing rear vehicle
                    OvertakeVehicle = RearVehicle
            }
else if     (
            RearVehicle and
            Lane (car) == LeftLane and
            ChangingLane (Car) == 'ChangingToRight' and
            Speed (RearVehicle) < Speed (Car) and
```

```
    Distance (RearVehicle, Car) < MMCO x ReactionDistance
    )
    {
        // Pulling back in front of rear vehicle
        OvertakeVehicle = RearVehicle;
    }
else
    {

        // Not overtaking
        OvertakeVehicle = null

    }
```

The Analyst/Planner

The Analyst/Planner receives its main input from the Sensors and, based on this and the rules which encapsulate the situation descriptions, it determines the actual tasks which the driver is performing and those which are expected.

It receives from the Navigation system a representation of where the vehicle currently is and it decides whether there is a need to communicate with the driver. Communication may occur because some information-providing device, such as the Navigation system, has indicated that a message should be sent, or it may occur because the Analyst/Planner decides that there is a threat to the driver's safety. An example of this might be the headway detector indicating an imminent collision. The Analyst/Planner is also responsible for assigning a priority to the messages it generates. These messages are then sent to the Dialogue Controller which schedules the incoming messages and, based on their priority and the current driver workload, decides whether and when to communicate them to the driver.

The Dialogue Controller

The Dialogue Controller has two major functions. First, to prevent several applications from communicating with the driver at the same time and, second, to adapt the attentional demands of interaction with GIDS to demands required by the traffic. In the initial GIDS prototype we have aimed primarily at preventing interference between different applications. Workload indicators have been incorporated, but they have only been estimated in a preliminary fashion and extensive empirical studies are required in order to fine-tune them.

However, a driver workload estimation algorithm has been incorporated (Piersma, 1991), that estimates driver workload on several resources and attention requirements of the situation as a function of driver characteristics (see Chapters 7 and 8).

6.7 Conclusion: the generic aspects

This section concludes the chapter by looking to the future and outlining how the current GIDS system can be extended. From the work which has been done so far it is clear that, although the Small World is in some respects a 'toy' world, it has been able to support a rich set of driving tasks. We believe we have shown that driving situations can actually be formalized, thus providing a computational model of the driving task. We see no major problems in increasing the set of situations thus defined so as to encompass most, if not all, situations likely to occur in the real world. Of critical importance for the system is that it has a sufficiently rich representation of the world in which the vehicle is moving. This representation was hand-coded in the case of the Small World. With the development of electronic maps for use with Navigation systems, or indeed the possibility that the local road topography could be beamed to a vehicle from roadside beacons, we believe that this representation could be produced and made available in future systems. Central to the philosophy of GIDS is that, as new information-providing devices and sensors become available, it should be possible to incorporate them into the system. We believe that this will indeed be possible and have designed the system so as to accommodate it. Once it is specified how the input from such sensors should be presented to GIDS, the intelligent components will be able to utilize it. Of particular interest will be fine-tuning the workload estimates and thus customizing the communication for the driver so as to further enhance safety. In particular, we hope that as a result of empirical studies the Analyst/Planner will be able to further augment the messages it passes to the Dialogue Controller with such information as (i) attentional demands of the message in relation to personal characteristics of the current driver (e.g., interference found with a specific secondary task) (ii) demands on vision (e.g., required reading time) (iii) required manual demand (e.g., expected response duration) (iv) required pedal demand (e.g., expected response duration) as induced by the driving task itself. The Dialogue Controller would use this additional information to determine more precisely whether and when messages with specified attentional and resource demands could safely be communicated. In this way, driver-GIDS interaction would be adapted even better to the continuously changing workload induced by the driving task. The current functions and workload adaptivity are subjects of the next chapter.

References

Aasman, J. (1988). Implementations of car-driver behaviour and psychological risk models. In J.A. Rothengatter & R.A. De Bruin (Eds.), *Road user behaviour: Theory and research* (pp. 106-108). Assen, The Netherlands: Van Gorcum.

Aasman, J., & Michon, J.A. (1991). Soar as an environment for driver behaviour modeling. In L.J.M. Mulder, F.J. Maarse, W.P.B. Sjouw, & A.E. Akkerman (Eds.), *Computers in psychology: Applications in education, research, and diagnostics* (pp. 219-226). Amsterdam: Swets & Zeitlinger.

Aasman, J., & Michon, J.A. (1992). Multitasking in driving. In J.A. Michon & A. Akyürek (Eds.), *Soar: A cognitive architecture in perspective* (pp. 81-108). Dordrecht, The Netherlands: Kluwer.

Akyürek, A. (1992). On a computational model of human planning. In J.A. Michon & A. Akyürek (Eds.), *Soar: A cognitive architecture in perspective* (pp. 81-108). Dordrecht, The Netherlands: Kluwer.

Allen, J.F., Hendler, J., & Tate, A. (Eds.). (1990). *Readings in planning.* San Mateo, CA: Morgan Kaufmann.

Boden, M.A. (1989). *Computer models of mind.* Cambridge: Cambridge University Press.

Cohen, R., & Perrault, C.R. (1979). Elements of a plan-based theory of speech acts. *Cognitive Science, 3,* 177-212.

De Velde Harsenhorst, J.J., & Lourens, P.F. (1987). *Classificatie van rijtaakfouten en analyse van rijtaakverichtingsparameters* [Classification of driving errors and analysis of driving performance parameters]. (Technical Report VK 87-17). Haren, The Netherlands: Traffic Research Centre, University of Groningen.

Goffman, E. (1959). *The presentation of self in everyday life.* Garden City, NY: Doubleday.

Kelly, J. (1990). *Philosophical problems of Artificial Intelligence.* PhD Thesis. Dublin, Eire: Department of Computer Science, University College Dublin.

McDermott, T.T., McLoughlin, H.B., & Toland, C. (1989). *Integrating user modelling to UIMS.* Technical Report UCD-CS-1989-0901. Dublin, Eire: Department of Computer Science, University College Dublin.

McKnight, A.J., & Adams, B.B. (1970). *Driver education task analysis. Volume I: Task descriptions.* Contract no. FH 11-7336, Final Report. Alexandria, VA: Human Resources Research Organization.

McLoughlin, H.B. (1987). Personae: Models of stereotypical behaviour. In R Reilly (Ed.), *Communication failure in dialogue and discourse* (pp. 233-241). Amsterdam: Elsevier Science Publishers.

Michon, J.A. (1985). A critical review of driver behavior models: What do we know, what should we do? In L.A. Evans & R.C. Schwing (Eds.), *Human behavior and traffic safety* (pp. 487-525). New York: Plenum Press.

Michon, J.A. (1987, September 2). Twenty-five years of road safety reseach. *Staatscourant, no. 168*, 4-6.

Michon, J.A. (1989). Explanatory pitfalls and rule-based driver models. *Accident Analysis and Prevention, 21*, 341-353.

Perault, C.R, & Allen, J.F. (1980). A plan-based analysis of indirect speech acts. *American Journal of Computational Linguistics, 6*, 167-182.

Piersma, E.H. (1991). Real time modelling of user workload. In Y. Quéinnec & F. Daniellou (Eds.), Designing for everyone (Vol. 2, pp. 1547-1549). London: Taylor and Francis.

Rich, E. (1980). Users are individuals: Individualizing user models. *International Journal of Man-Machine Studies, 18*, 199-214.

Schank, R.C, & Abelson, R.P. (1977). *Scripts, plans, goals, and understanding: An inquiry into human knowledge structures.* Hillsdale, NJ: Erlbaum.

Smiley, A., & Michon, J.A. (1989). *Conceptual framework for generic intelligent driver support.* Deliverable Report DRIVE 1041/GIDS-GEN01. Haren, The Netherlands: Traffic Research Centre, University of Groningen.

Van Winsum, W. (1991). *Cognitive and normative models of čar driving.* Deliverable Report DRIVE 1041/GIDS-DIA03. Haren, The Netherlands: Traffic Research Centre, University of Groningen.

Webster, E., Toland, C., & McLoughlin, H.B. (1990). *Task analysis for GIDS situations.* Research Report DRIVE/GIDS-DIA. Dublin, Eire: Department of Computer Science, University College Dublin.

Chapter 7
GIDS functions

Willem B. Verwey, Håkan Alm, John A. Groeger, Wiel H. Janssen, Marja J. Kuiken, Jan Maarten Schraagen, Josef Schumann, Wim van Winsum, Heinz Wontorra

7.0 Chapter outline

From the general features of driver support discussed in Chapters 4 and 5 we now proceed to a more detailed exposition of the specific functions selected for each of the functional levels we have distinguished there: planning and navigation, manoeuvring, and control. This chapter provides a systematic treatment of guidelines for each of these major functions in GIDS, also frequently referred to as *applications*. Subsequently, in Chapter 8, the implementation of these functions in the GIDS system will be addressed.

The functions described include navigation support (Section 7.1), collision avoidance support (Section 7.2), control support through active control devices (Section 7.3), adaptive support of novice and inexperienced drivers (Section 7.4), and the GIDS calibrator (Section 7.5). The other – extrinsic – functions in GIDS, the car telephone and stereo system, have not been subjected to theoretical and experimental evaluation. These are treated as fixed elements in the GIDS architecture and will be discussed in Chapter 8. The final section (Section 7.6) describes how the various functions should be integrated into one coherent system in which all functions are adapted to human capacities without interfering with each other or with the driving task.

7.1 Navigation support

When travelling in a partly or totally unfamiliar environment, drivers require information in order to reach their destination. In a highly familiar environment it may happen that they do not know which route is quickest, given the current traffic situation. In both cases drivers will benefit from an intelligent support system that could guide them to their destination via the quickest route. The principal aim of a navigation system is to provide these functions. But at the same time the system should be designed in such a way that driving safety is not jeopardized by presenting the information in a complex way. This section briefly reviews basic questions on the ergonomics of in-vehicle navigation systems and it concludes with a number of specifications.

Basic questions

(a) Criteria for route selection

Sophisticated navigation systems in the car usually select a route between an origin and a destination. For this route selection a number of different strategies can be applied. Optimization on distance is the most frequently used alternative. In that case, distance is the only attribute of the route taken into consideration by the route-selection algorithm.

The way the route is selected and the attributes taken into account during the route-selection process are of great importance. In order to ensure a high level of driver acceptance, the route-selection logic should match the decision logic of the user of the system. It is assumed that system acceptance is promoted if drivers are guided along routes they would have selected themselves if they had possessed full knowledge of all available routes. In order to choose the right kind of attributes of routes and to establish weights of these attributes for the design of a route-selection algorithm, it is necessary to know the drivers' needs. Needs, in this context, relate to the reasons why car drivers prefer some routes over other routes. Van Winsum (1989) has reviewed a large number of studies on route choice by car drivers. None of these studies has considered the decision rules of car drivers in sufficient detail for our present purposes. Nevertheless, some factors do stand out as determinants of route choice. Time savings and distance are most frequently reported as relevant criteria. Safety of the route is found to be important in some studies but not in others. The same applies to criteria relating to congestion and self-pacing, that is, the possibility to determine one's preferred speed independent of the prevailing circumstances. Scenery seems to be important in non-business-related trips but not in business-related trips. Significantly, cost appears to be unimportant as a route-choice criterion.

Thus, travel time, distance, congestion, safety, and the possibility to keep moving appear to be important. A problem, however, is that these criteria are not mutually independent.

In order to establish the weightings of route criteria by car drivers in their route choice, a decision-analysis experiment based on the multi-attribute utility (MAU) method was performed (Van Winsum, 1990). In this study nine attributes were considered:

— time loss caused by giving right of way;
— time loss caused by traffic lights and queues;
— number of lanes and maximum speed (road type);
— speed adaptation caused by other traffic;
— number of turns;
— predictability of journey time;
— attention required;
— scenery;
— distance.

Between-subject factors were gender, age, education, trip motive and driving experience. It was found that none of the latter factors influenced the subjective importance of route choice criteria in a systematic way. Further, it was found that the preference for a route was strongly determined by 'distance', 'time loss caused by traffic lights and queues', 'time loss caused by giving right of way', 'road type' and 'predictability of journey time'. The other attributes proved to be less important determinants of route choice. 'Predictability of journey time' was excluded from further analysis because it is not a clearly measurable objective attribute. Route selection algorithms were simulated by computing the effect of inclusion of an attribute on the correlation between the predicted route preference and the actual route preference. Route preference was predicted by summing the products of the weights and the values of the attributes.

It was found that the attributes 'time loss caused by giving right of way', 'time loss caused by traffic lights and queues', 'road type' and 'distance', with the exclusion of all other attributes, resulted in prediction of routes that were the actually most preferred route for 82% of the subjects. Only 14% of the predicted routes were second best according to the subjects. Worst routes were never predicted and only 4% of the predicted routes were second worst according to the subjects. If distance was used as the only prediction criterion, as in a number of conventional route selection algorithms, 41% of the predicted routes were first choices by the car drivers. The fact that this figure conforms with the literature (Koning & Bovy, 1980) strongly suggests that people often choose routes which are sub-optimal on distance because criteria other than distance are also taken into account. Since there were no significant effects of gender, age, trip type, educational level, and driving experience, the attributes of routes and their weights that are to be used for route selection can be the same for all drivers.

.(b) Sensory modalities to be considered

A human being has five input channels to the internal and external environment: visual, auditory, tactile, kinaesthetic, and olfactory/gustatory. Drivers may receive messages through any channel in isolation, or through a combination of two or more channels.

For a navigation system the most important channels are the visual and the auditory. Driving is largely dominated by visual information (Rockwell, 1972; Verwey, 1991). For this reason it may be advisable to avoid making the visual scene even more complex by introducing additional systems displaying visual signals or messages and to favour, instead, the use of auditory information.

Several research efforts have been made to compare the effectiveness of auditory information with that of visual inputs. Streeter, Vitello, and Wonsiewicz (1985) concluded that carefully designed auditory instructions presented on tape appear to have an advantage over customized route maps. Subjects who used oral instructions to reach a particular destination drove shorter distances and made fewer errors than subjects who used the route maps. Somewhat surprisingly, Streeter et al. also found that the combination of route maps and oral information led to poorer performance compared with oral information only. Unfortunately, they only carried out statistical tests on a subset of the navigation errors, leaving out those they did not consider "true" errors. Verwey and Janssen (1989) also found that auditory information can be superior to visual. They concluded that auditory information seems to be best for drivers who are unfamiliar with navigation systems. However, with practice the positive features of visual information may become more prominent. Davis and Schmandt (1989) also argue for the superiority of oral information and consider how to overcome problems associated with the use of spoken messages.

Verwey (1989, 1992) reports evidence, obtained in two laboratory studies, for the notion that people recode visual/spatial information (i.e., arrows) into a verbal memory code. On this basis, using oral information to guide a driver along a route seems quite promising, especially if the traffic situation is demanding and requires the driver's visual attention to a very high degree.

Nevertheless, it may be premature to dismiss the use of visual information. Visual information processing has the advantage of being self-paced, as was pointed out by Verwey and Janssen (1989). This means that, so long as the information remains on display, drivers can quickly pick it up whenever they consider it safe to do so. In other words, visual navigation information may be useful if it is kept very simple, if it is located near the driver's forward line of sight and if it remains displayed until the driver has used it. A study by Alm and Berlin (1991) has shown that a relatively large proportion of drivers need repetition of an oral route guidance message, even if it is given only a few seconds before the intersection to which it applies. This brings out another problem with oral messages: they take time, to initiate, to perceive, and to understand. Altogether, it is recommended to use a simple visual message as a memory aid to oral messages. This eliminates the need for repeated presentation of the oral information.

(c) Type of information to be presented
Given the present assumption that both oral and visual information should be used, a question arises about the content of the information. A navigation system must, in some way, refer to the environment. This can be achieved by giving an oral or visual description of (part of) the environment, or by pointing at some part of the environment. In both cases designers must know what the system should indicate and how that should be accomplished. A system adapted to the driver's needs and capabilities will match its references to the world with people's perceptions of the world. Thus, we need to know more about drivers' mental models of the driving environment. On the basis of a literature review which included studies that compared some form of navigation system with the use of a conventional map, Schraagen (1991) provided the following guidelines for information presentation: (1) map-like displays should not be used in navigation systems; (2) street names should not be used; (3) simple left/right instructions, whether orally or visually presented, lead to the largest reduction in errors and should therefore be used in navigation systems.

Two empirical studies within the GIDS project have specifically dealt with this issue (Alm, 1990; Schraagen, 1990). It was found that drivers who are familiar with the environment rely on landmarks in their internal representation of routes. Instead of using distance, such subjects often use landmarks to define a choice point. This has also been reported by Davis and Schmandt (1989). In particular, traffic lights, traffic and orientation signs, fuel stations, and shops were frequently used. These landmarks can be used to indicate a route more clearly. It was also found that subjects who relied heavily on street names while navigating through unfamiliar environments, by having memorized routes on maps, made more navigation errors than subjects who attended to landmarks on the maps (Schraagen, 1990).

In the present conception of the GIDS navigation system, simple left/right instructions are presented both orally and visually, and landmarks such as traffic lights are used to clarify these instructions.

(d) Amount of information to be presented
A navigation system should not present too much information. This may, in the worst case, lead to confusion and distraction or, less seriously in terms of safety, to disregard of relevant information.

In order to determine the optimal amount of information required from an orally based navigation system, a field study was carried out within the GIDS project (Alm & Berlin, 1991). Twenty-four subjects, randomly assigned to three experimental conditions, were given the task of driving to an unknown destination. Each group received a different amount of information. The results of the study indicate that information about one or two choice points at a time seems to be optimal, and that the navigation system should vary the amount of information depending upon the driving time between choice points. That is, the amount of information given by the navigation system should (ideally) be dependent upon the structure of the driving environment and the way the person drives in order to minimize memory errors. One important aspect of the environment is the driving time between sections where a decision must be made. This time varies, of course, as a function of the traffic situation and the driver's time demands.

In another study in the GIDS project (Schraagen, 1991) it was found that three navigation instructions presented simultaneously resulted in more navigation errors than single instructions presented in the form of arrows. Therefore, whether presented visually or auditorily, navigation instructions should be limited to one or two instructions. In the GIDS navigation system, at no time are more than two instructions presented. In fact, the only time two instructions are presented simultaneously, is when the driving time between choice points is less than 10 s.

Conclusions and recommendations

The literature surveys and the empirical studies discussed so far lead to the following recommendations for a navigation system:

- Route selection algorithms should make use of the following criteria: waiting time for ceding right of way, waiting time for traffic lights, road type (with associated speed limit), and distance;
- The system should be able to present both oral and visual information: this will give the driver the opportunity to select the source of information depending upon the demands of the road and traffic situation and to confirm the last oral instruction;
- The content (meaning) of the oral and visual messages should be the same;
- The visual messages should consist of arrows, pointing in the required direction of travel. The oral messages should consist of simple instructions such as "turn left" and "turn right". Simplified representations of landmarks should be presented in colour. No map-like displays should be used;
- If it is necessary to clarify a route or an exit from one road to another, landmarks such as traffic lights, traffic and orientation signs, shops, and fuel stations should be used. On motorways clarification by reference to traffic signs is recommended;
- A navigation message should preferably concern one choice point at a time, but if the driving time between them is less than 10 s two instructions may be presented.

7.2 Collision avoidance support

The purpose of a collision avoidance system (CAS) is to reduce the frequency of collisions and near-collisions. A CAS which operates in the longitudinal dimension, that is, in car-following situations, would thus have to eliminate the occurrence of very short headways between vehicles.

There are two main issues to be resolved before a well-functioning CAS can be designed. First, there is the issue of the criterion for activating such a system. Second, there is the issue of what action should be performed after the system has been triggered. This section will discuss both issues; it will briefly present research findings bearing on each, and it will conclude with specifications for a CAS as part of a prototype GIDS system.

Criteria for CAS activation

Three simple criteria for CAS system activation may be distinguished a priori:

(1) A criterion that is fixed in terms of time; that is, the unconditional activation of the system by any obstacle within a certain time range of the CAS vehicle. Because this 'brick wall' criterion does not take into account any movements of the obstacle itself, it may lead to system activation in cases that would never in practice develop into collision configurations.

(2) A criterion that is defined in terms of the existence of a momentary collision configuration, that is, a combination of positions and velocities from which a collision would ensue if these parameters remained unchanged over the next, say, 4 s. Intuitively, this 'time-to-collision' (TTC) criterion could be expected to result in smaller numbers of false alarms than the 'brick wall' criterion, while capturing all configurations that will eventually lead to collisions and some that will not.

(3) A conditional criterion, that is, the existence of a configuration that would turn into an actual collision configuration if one of the parameters were to be changed suddenly, in particular, by the sudden application of full braking power by the vehicle in front. It may be noted that this 'worst case' criterion is more conservative than the TTC-criterion, but considerably less so than the 'brick wall' criterion (because it takes into account that the preceding vehicle is moving).

Form and allocation of evasive actions

Apart from the criterion problem there is the issue of what action will have to be performed once the criterion has been met. In particular it will have to be decided what mode of allocation of responsibility for action should be chosen so as to optimally divide the burden of further action between the driver and the automated system. The following are conceivable solutions:

– Displaying the relevant variable to the driver in a continuous manner.
– Warning the driver when the relevant variable (distance, TTC) approaches the set criterion value.
– Initiating an appropriate movement of the accelerator pedal which drivers cannot neglect. The obvious advantage of such active controls, e.g. the accelerator rising up under the driver's foot when deceleration is urgently required, is that they reduce driver reaction times in those situations. This makes them attractive, a priori, as part of a CAS. One might uphold the view that ultimate control over this type of device should be left to the driver, that is, that the driver should be permitted to overrule the active control if this was deemed necessary. Alternatively, one might opt for the complete elimination of the driver as a decision maker in this type of action. This leads to the last alternative, total system take-over;
– Total system take-over.

Experimental research on *what* and *how* in a CAS

A number of the considerations on CAS design that were discussed in Section 4.2 have been put to the test using driving simulators (Janssen & Nilsson, 1990; Nilsson, Alm, & Janssen, 1991). In these experiments subjects driving on a simulated two-lane road were presented with traffic in front of them and their behaviour was observed with and without a CAS in their vehicle. Behaviour was defined in a general sense here, that is, both in terms of what drivers did when closing in on a leading vehicle and in terms of their general driving style.

The tested systems used combinations of some activation criterion (either a 4 s time-to-collision or a 'worst case' criterion) and some system action. The action could be a warning light or buzzer activated inside the CAS vehicle and, in one particular form, an active control. The simulator's accelerator pedal in this case was activated by a 25 N increase in pedal force whenever and for as long as the criterion was met. Thus, it would rise under the driver's foot – thereby reducing speed – although it remained possible for the driver to depress the pedal and increase speed.

CAS effects on car following were evaluated by noting the amount of time during which subjects were less than 1 s from the leading vehicle. There is evidence in the literature that headways below 1 s are predictive of future accident involvement (e.g., Evans & Wasielewski, 1983). Therefore a reduction in this behaviour would represent a real safety gain. All tested CAS systems showed positive effects in this respect, that is, a reduction in short headways (see Fig. 7.1). This gain was largest for the CAS in which the counter-action of the accelerator was coupled to a 4 s time-to-collision criterion. Whereas approximately 5 percent of time headways were below 1 s without CAS, this dropped to approximately 2 percent with this particular CAS activation criterion.

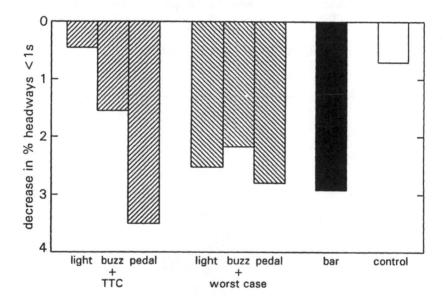

Figure 7.1 Safety gains of CAS systems

Side-effects of CAS

Apart from an overall positive effect on following performance per se, several more general effects on driving style were found with most CAS systems tested in this study. First, overall driving speed with a CAS increased by some 2 to 4 percent (see Fig. 7.2). Second, there was an increase in speed variability in most systems. Third, there was a general tendency for driving in the left lane of the road to occur more often with a CAS; on a two-lane road, as used in the experiments, this may of course amount to exchanging one risk for another (see Fig. 7.3).

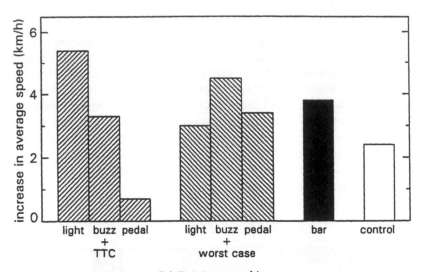

Figure 7.2 Driving speed increase

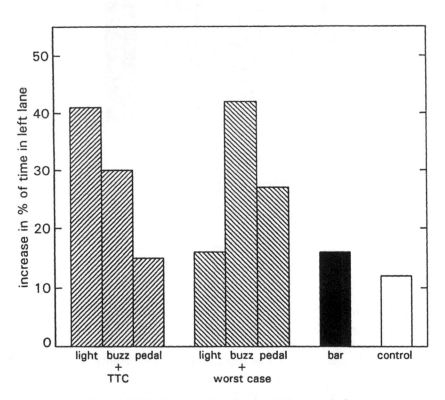

Figure 7.3 Collision with a lead vehicle in right lane

These side effects more or less offset the safety gain obtained in car following performance using CAS, although it is impossible to say what the net effect in terms of safety would be in specific cases. Fortunately, however, there appears to be one CAS which, while producing a reduction in short headways, suffered little from counterproductive effects in overall speed, speed variability, and/or the tendency to drive in the left lane. This was the 'TTC + Active Accelerator Pedal' system. The natural reduction in speed offered by this device whenever TTC dropped below its criterion was indeed accepted by drivers who did not, for instance, move over to the left lane to compensate for the incurred time loss.

Conclusions and recommendations

On the basis of the reasoning and the research results described in this section it may be concluded that the longitudinal collision avoidance component to be part of GIDS should comprise the following elements:

–   A time-to-collision (TTC) criterion for system triggering which is of the order of 4 s;
–   System action, after triggering, in the form of an increased accelerator pedal counter force, the action being overrulable by the driver (by applying an extra force).

7.3 Active control devices

In GIDS, two kinds of active controls have been studied, the steering wheel and the accelerator pedal. This section first gives a brief summary of the available literature on active controls and then reports some studies on the application of active controls in GIDS. The section ends with conclusions and recommendations.

Review of literature

Active control devices may be used to convey relevant information to the driver and they therefore serve concurrently as proprioceptive-tactile displays (Rühmann, 1981). An example was discussed in the previous section. The application of active controls can be described from an information processing perspective within the concept of stimulus-response compatibility (Sheridan & Ferrell, 1974; Wickens, 1984), since the receptors of the information synchronously serve as effectors for the action indicated by the stimulus information. High stimulus-response compatibility results in reduced workload, faster response times, and fewer errors (Sanders & McCormick, 1987).

Active control devices providing proprioceptive-tactile information have been developed mainly for aircraft control (Jagacinski, Flach, & Gilson, 1983; Hosman & Van der Vaart, 1988), but there have been sporadic attempts to use them in automobiles. Fenton (1966) implemented a manual control unit for headway support, replacing the normal steering wheel and pedals by a control stick (Rule & Fenton, 1972). Panik (1984) carried out an empirical study on the effectiveness of an active accelerator pedal with programmable force feedback for car-following behaviour. Only recently, through the implementation of computer technology in automobiles (e.g., the design of *drive-by-wire* steering), has active control technology also been under consideration in motor vehicle control (Hess & Modjtahedzadeh, 1990). In the GIDS system, active control devices have been incorporated as-an integral element of the overall system to support longitudinal (active accelerator pedal) and lateral (active steering wheel) control.

Experiments on basic signal intensities

As a basis for the implementation of an active steering wheel, the necessary signal intensity that can be perceived unambiguously at the steering wheel has been determined experimentally. The signal that served as a proprioceptive-tactile information cue to the driver consisted of a steering wheel torque shift, that is, a short rectangular positive torque pulse of 0.5 s superimposed upon the normal steering wheel torque (Färber, Br. Färber, Godthelp, & Schumann, 1990). Brief signals were chosen in order not to disturb or distract the drivers' internal model of their car, a point also stressed by Kelley (1968). In a psychophysical experiment, *just noticeable differences* for this torque shift were determined on the basis of different steering wheel torques. This resulted in a recommended signal intensity of approximately 1.2 Nm (Schumann, Färber, & Wontorra, 1991).

To examine the effectiveness of proprioceptive-tactile feedback via an active accelerator pedal, a laboratory study was carried out in a fixed-base driving simulator (Godthelp, 1990). Different force feedback characteristics were compared in a tracking experiment. The results showed that tracking errors are significantly smaller in conditions with active accelerator pedal support, suggesting that this support may effectively reduce workload. Based on the results of this experiment, a servo-controlled pedal with a maximum force level of 350 N was recommended.

Active control devices for specific driving tasks

(a) Lane change and speed regulation on straight roads
The results of these preliminary experiments were used in a series of driving simulator studies, in which different active accelerator pedal and steering wheel support strategies were tested and compared for different driving situations and tasks. The central question in the first experiment was whether an active steering wheel can be

implemented to transmit relevant proprioceptive-tactile information cues for lateral control. Therefore, the use of proprioceptive-tactile information signals was examined in a lane-keeping task (Schumann, Godthelp, Färber, & Wontorra, in press). The driver had to initiate a lane change manoeuvre, the execution of which can be described as open-loop control behaviour (see Section 4.4). During this manoeuvre a warning could occur informing the driver to stay in the right lane, resulting in a transfer from open-loop control to closed-loop control. Warnings were given at a certain time-based distance (2 s) from the road centre-line (operationalized by a time-to-line-crossing value, TLC) and presented to the driver by means of one of two discrete steering wheel torque changes (i.e., a short vibrating torque shift or a short steady torque shift). The results showed that proprioceptive-tactile warning signals are appropriate to interrupt an intended lane-change manoeuvre and cause the driver to stay in the right lane.

In addition, an active accelerator pedal study was carried out. It focused on the question as to which force feedback characteristics would be most appropriate for driver support in speed regulation. To examine the use of an active accelerator pedal for speed control, subjects performed a speed adaptation task on a straight road approaching a speed limit area (Godthelp & Schumann, 1991). Different pedal-force feedback characteristics were compared, the force being related to the error in speed or accelerator pedal position. Driver performance was evaluated in conditions with normal and with restricted view of the speedometer. The results showed that force feedback from the accelerator pedal can serve as a useful means of reducing speed errors. Comparing the different accelerator pedal configurations used in this study, it appeared that force feedback related to some speed-error criterion leads to fewest speed errors. Results from driver speedometer observations indicated that this type of control support may also affect looking behaviour, since speed error feedback allows the driver to reduce the number of speedometer glances. This finding confirmed the idea that force feedback from the accelerator pedal may improve the quality of the driver-car interface from the point of view of workload.

In a next experimental step, the question was whether there is interference when both active control devices are implemented and warning signals appear simultaneously on both controls. The task for the driver consisted of a combination of the above speed-adaptation experiment and a lane-changing task. Four different signal combinations were compared, that is, short or longer vibration of the steering wheel, and short vibration or force feedback related to the speed error via the accelerator pedal. With one exception, the results showed no negative mutual interference. The longitudinal control task (speed adaptation) in particular was performed very well. The exception concerned the proper execution of the lateral control task (lane changing) when similar vibrating signals were presented simultaneously on the two active controls. The reason that only lane-changing performance was affected may be that the steering task demands more attention than the longitudinal control task in which the foot has only to be removed from the accelerator pedal. With this exception, it appears that both active control devices may be employed simultaneously.

(b) Curve driving

A further experiment was conducted to test the applicability of lateral and longitudinal control support in critical curve driving by means of active steering wheel feedback and/or active accelerator pedal feedback. Criticality was established in two ways: (a) in terms of curve geometry (clothoids of constant length with an ingoing and an out-going section, but with different curvature gradients), and (b) by means of brief (1 s) occlusion of vision. The accelerator pedal was set up for force feedback related to some fixed speed error (difference between actual speed and target speed, calculated according to the curve's ideal speed profile). The control condition involved an autopilot, that is, speed was generated by the system itself, using the curve speed profile. Steering wheel feedback was set up as a single short steady torque shift towards the desired direction, triggered at two different time-based, critical thresholds relative to the roadside. The critical thresholds were operationalized as time-to-line-crossing (TLC) values of 0.75 and 1.25 s, respectively. The data showed that speed variances were not negligible, neither as effects of the accelerator pedal feedback mode, nor as between-subjects effects for the speed error guidance mode. For that reason, lateral tracking performance was analyzed separately for the two accelerator pedal feedback conditions. In the autopilot condition, steering performance deteriorated if the visual flow was interrupted by occlusion. The two TLC triggering thresholds did equally well and both were better than the unsupported steering wheel. The results also reflected no differences between the various curves. This is encouraging in that the applicability of steering support is apparently not restricted to special curves. For the speed support condition, the results indicated an effect of occlusion, showing the same pattern as under the autopilot condition. However, contrary to the autopilot findings, there was no positive influence of steering wheel warning on lateral control performance. This may be attributable to the secondary task of speed control. That is, the accelerator pedal may have been a distractor for the lateral tracking task. One possible counter-measure against this distraction might be a slightly increased steering wheel signal (see Färber, Godthelp, Schumann, & Wontorra, 1991).

Interaction of warnings in critical driving situations

Based on the findings of the former experiments, a final simulator study was carried out in the Daimler-Benz driving simulator in Berlin. One question concerned the additional impact of kineasthetic sensations from the car (e.g., lateral and longitudinal acceleration) on the effective perception and processing of proprioceptive-tactile information signals. The goal of this investigation was to verify the findings of the earlier studies in quasi-realistic critical traffic situations. These critical situations were *rare events* – in so far as this can be established in a simulator study – in a series of quite normal driving situations occurring during a normal journey. For three prototypical, critical situations the applicability range of accelerator pedal and/or steering wheel control support was tested:

(a) entering a curve of non-constant curvature (clothoid) that could not be previewed over its total length; the initial speed was higher than the design speed for this particular curve;
(b) an unpredictable braking manoeuvre of a leading vehicle;
(c) an overtaking manoeuvre that had to be aborted because of undetected oncoming traffic.

For the curve driving situation an accelerator pedal force feedback was used, based on speed error (actual speed minus target speed, calculated according to the curve's optimal speed profile), and this was tested with two different steering wheel warning systems. The steering wheel feedback consisted of single torque shifts, triggered by reference to a time-critical and/or a lateral-distance-critical threshold.

Using both warning criteria, the discrete steering wheel warning did not lead to a significant improvement of lateral control performance. Since the signals were presented during the closed-loop control phase, this interfered with the control strategy of the driver. In this phase the driver mainly relies on optical and kinesthetic cues to perform the driving task. The accelerator pedal force feedback did lead to an improvement of the longitudinal control performance.

The braking manoeuvre was supported by force feedback on the accelerator pedal only. As soon as the distance from the leading vehicle fell below a so-called worst-case distance (see Section 7.2), a continuous force increment was superimposed on the normal accelerator pedal force. When the actual distance fell below the distance corresponding to a time-to-collision (TTC) of 4 s, this force increment reached its maximum value. Rear-end collisions were reduced significantly with this accelerator support. However, since in some situations the driver can be disconnected from 'the accelerator display' (e.g., when using the brake pedal) an additional warning (e.g., head-up display) in critical driving situations (e.g., a TTC of 4 s) should be presented.

For the overtaking manoeuvre, two steering wheel warning modes were combined with an accelerator pedal feedback. The steering wheel warnings were a single torque shift to the right or a vibrating signal on the wheel, both indicating that the driver should stay in the right lane. Both signals were triggered as soon as the vehicle's left bumper passed the centre line of the road. When steering back to the right lane, the accelerator pedal feedback made the subject aware of the distance from the leading vehicle. The feedback force was calculated according to the above-mentioned criteria. The results showed a supportive effect of the active steering wheel in interrupting the intended overtaking manoeuvre. The signal intensity of the short proprioceptive-tactile warning should be in the region of 1.5 Nm. The accelerator pedal warning resulted in an additional supportive gain. After the overtaking manoeuvre had been aborted, the minimal distances to the leading vehicle were increased. The results for curve driving and for prevention of an intended overtaking manoeuvre were confirmed in an experiment carried out on a closed circuit (see Färber, Naab, & Schumann, 1991).

Conclusions and recommendations

On the basis of the results of these experiments it is reasonable to divide the recommendations into three parts: those concerning the active steering wheel, the active accelerator pedal, and the possible interactions of both active control devices, respectively.

(a) Active steering wheel:
- During open-loop control, warning signals on the steering wheel should be brief and discrete in order not to disturb the driver's internal model of the normal steering characteristic;
- Two signal forms are conceivable: unspecific vibrating torque shifts and specific, directional steady torque shifts. The choice between these signals depends on the control manoeuvre during which the signals will be presented;
- To interrupt an intended lane-change manoeuvre (open-loop control mode) an unspecific vibrating signal is sufficient;
- For curve negotiation in the open loop control phase, a specific, directional steady torque shift may be appropriate. More research is needed for this application;
- During closed-loop lane keeping (e.g., traversing a curve), discrete force signals should be avoided. They interfere with the driver's intentions and closed-loop control behaviour;
- In some situations (e.g., leaving the road) a continuous feedback might be helpful, even during closed-loop control. This should be tested in more detail. Such a continuous feedback presupposes, however, knowledge of all relevant driving parameters (e.g., the exact course of the road, other traffic) and the driver's normal closed-loop behaviour;
- The steering wheel support should be activated, for discrete warnings as well as for continuous feedback, at the latest safe moment;
- Recommended signal intensities are of the order of 1.5 Nm.

(b) Active accelerator pedal:

- Since the regular feedback mechanisms for longitudinal control are not so directly related to drivers' actions, their internal model is relatively fuzzy. The danger of disturbing this internal model by additional continuous force feedback is therefore less critical;
- For speed regulation a continuous force feedback via the accelerator pedal, that is related to speed error calculated by reference to some normative value, is recommended;
- Pedal forces should be contained within an upper boundary of 250 N and a lower boundary of 10 N;
- For collision avoidance a discrete signal force is recommended; since the driver is sometimes disconnected from the 'accelerator display' an additional warning (e.g., head-up display) should be presented.

(c) Interaction of both active control devices:

- A tuned, successive use of both active control devices may be useful during specific driving manoeuvres involving lateral and longitudinal control (e.g., interrupting an overtaking manoeuvre);
- A simultaneous presentation of discrete warnings via the active control devices may disrupt the steering task. Both discrete signals attract the driver's attention and impair the more complex anticipated control action, that is, steering. Therefore such a simultaneous presentation of discrete warnings via the active control devices should be avoided.

7.4 Adaptive support

Since feedback on performance is available to drivers only under exceptional circumstances, one of the fundamental building blocks of learning – knowledge of results – is normally denied to drivers as a means of improving their performance. It has been argued elsewhere (Groeger, 1991) that this alone is sufficient to reduce drivers' capacity to meet the requirements of their task, but that providing drivers with adaptive support (i.e., support which is appropriate for a particular situation and a particular driver) might well help to overcome these problems.

Aim of the function

The objective of the studies on adaptive support was to develop a means of augmenting the support or information normally available to drivers, in order to enhance their opportunities for learning and adaptation. Success in reaching this objective is ultimately dependent on identifying the support needs of different drivers in different situations and determining how that support should be delivered.

Empirical studies

Three different empirical approaches were adopted in order to address the issue of what drivers' support needs are: surveys, observation, and measurement of actual driver behaviour. The major outcomes of each approach are described below.

(a) Questionnaire on the behavioural antecedents of accidents
Findings from various studies associate age and traffic experience with differential involvement in accidents. There is, however, still a lack of knowledge on performance differences of drivers with differing levels of experience, precisely the information required in order to develop an adaptive support system. A questionnaire was constructed on the basis of a literature survey and information collected from in-depth interviews with some 30 drivers (see Groeger et al., 1990). Nearly 1000 drivers with differing levels of traffic experience were asked to describe a maximum of six accidents or near-accidents in which they had been involved, as well as the errors they had made prior to these accidents or near-accidents. The drivers reported in total 1786 accidents and near accidents, and 1670 behavioural errors made prior to these accidents. Inexperienced drivers reported relatively more accidents related to reversing, parking, and negotiating bends. These incidents also occurred more often during the first five years of driving. Experienced and very experienced drivers reported more rear-end collisions, accidents due to a sudden obstacle on the road and, to a lesser degree, accidents while negotiating a junction. The three most frequently reported errors were: not looking in the appropriate direction, incorrect interpretation of the traffic situation, and excessive speed. Observation of the traffic environment and making decisions or judgements were reported as two of the main sources of errors. Table 7.1 shows the relative contribution of different errors to particular accident situations.

There were no significant differences between different groups of drivers in their views on the contributory errors in certain types of accidents. However, there were differences between the groups when asked about the likelihood of occurrence of a series of behavioural errors. Errors relating to motor skills (i.e., vehicle control) were mentioned more often by inexperienced drivers. Errors relating to cognitive skills (estimation of speed, time and distances, knowledge of traffic rules) were also more often mentioned by inexperienced drivers. This corresponds to the greater cognitive load found among inexperienced drivers (Verwey, 1991). Experienced and very experienced drivers more frequently indicated that they made errors relating to a momentary state (e.g., being hurried, bored, careless, or drowsy). Drivers frequently reported that driving at excessive speed, either by themselves or by other car drivers, was the primary cause of accidents. In this respect they specifically mentioned accidents or near-accidents when taking a bend and rear-end collisions. Maintaining a specific course was reported to cause few problems by fully-licensed drivers.

Table 7.1 Errors most frequently reported within accident situations

Error Situation	1	2	3	4	5	6	7	8	9	10	11	12	13	14	15	16
Reversing	**		**					**								
Rear-end collision	**			**	*				**							**
Parking	**	**	**													**
Crossing junction	**			*		**										
Sudden obstacles	**															
Curve in the road	**				**											
Turning left	**					**										
Changing lane	**	**					*									
Overtaking	**	**		**			*									
Turning right	**	**														
Moving off	**															*
Roundabout	**															
Merging																
Other	**	**	*		*											**

Legend

*   Proportion greater than expected by chance (p<0.05)
**   Proportion greater than expected by chance (p<0.01)

List of errors

1   Did not look in the appropriate direction
2   Incorrect interpretation of situation
3   Steering error
4   Wrong estimation of speed of other traffic
5   Speed too high
6   Did not see traffic from the right
7   Did not remember blind spot in the mirror
8   Did not use mirror
9   Reaction too slow
10   Braking error
11   Wrong anticipation of situation
12   Did not see traffic from left
13   Wrong position on the road
14   Insufficient knowledge of traffic regulations
15   Error with gears
16   Other errors

(b) Task performance in Small World situations

In a subsequent study by Kuiken et al. (1991), quantitative measurements of driving performance (speed choice, lateral position, and steering movements) were obtained while drivers drove an instrumented vehicle through real-world versions of 'Small World' scenarios (e.g., negotiating curves or intersections, driving ahead on a straight road, negotiating a roundabout). The objective of the experiment was to see whether significant, observable differences could be found in the performance of groups of drivers with different levels of experience, driving a test route of 45 km, roughly a 1.5 hour drive. Drivers with different amounts of traffic experience were tested under comparable conditions, but few statistically reliable differences emerged. Thus it appeared that these groups of drivers performed similarly although they differed considerably in their experience within the traffic environment. Because people's performance on the indices used was so heterogeneous, it effectively means that support needs cannot be determined simply from knowledge of how long a driver has held a licence, or what distance they have driven during that time.

(c) Oral support provided during driver training
As part of an attempt to study the support needs of a homogenous and potentially high-risk group, an extensive analysis was made of the oral support received by a sample of drivers when undergoing training (Groeger et al., 1990). The driving lessons received by 20 British trainees were video-recorded and analyzed, from their first lesson (at which point they had no prior experience as motorists) to the point at which they took the state driving examination. Given an acceptable level of performance in this test, the British pupil is subsequently entitled to drive without being accompanied by another (qualified) driver. Unusually for Britain, the trainees in this experiment did not drive except under the supervision of their instructor and thus a full record of their early driving experience became available. In all, over 600 hours of video recordings were analyzed and some 40,000 comments made by instructors were put into a database. Comments were coded in such a way that the situation in which each comment was made together with its detailed contents (i.e., component activities referred to) could be reconstructed. These were later used to inform the design of the support system. For 10 of the pupils, every traffic situation they encountered during every lesson was also logged.

Table 7.2 summarizes some of the findings that are reported in detail in Groeger and Grande (1991), showing the number of times each Small World manoeuvre was carried out before a driver reached a test criterion, and how frequently the pupil's driving in that situation elicited comment from the instructor. The table also helps to identify those activities which are likely to require comment in each situation, and whether the level of support received changes as the pupil's experience grows (see suffix letters, e.g., e: even throughout course; d: decreasing throughout course; i: increasing throughout course).

Table 7.2 Manoeuvres carried out and frequency of comment during driver training (excluding simple driving ahead)

Legend

Supporting comments
=: support provision remains stable for that manoeuvre
dec: support provision decreases for that manoeuvre
e: support for this activity "even" throughout course
d: support for this activity "decreasing" throughout course
i: support for this activity "increasing" throughout course

Activity commented upon
1. Use of accelerator, brake, clutch
2. Steering
3. Observation
4. Judgement
5. Speed
6. Road positioning
7. Use of signals (indicators)

	TOTAL	%SUPP	Req	1 con	2 ste	3 obs	4 jud	5 spe	6 pos	7 sig
Moving off	152.30	57	dec	34d	12d	24d	5d	5e	12d	8d
Ahead near controls*	131.30	27	dec	21d	1e	27d	11e	24d	14d	2e
Ahead near others	164.00	91	=	13d	1d	19d	14e	7d	45d	1d
Overtaking	50.00	82	=	11e	8e	11e	24e	18d	20e	8e
Lane change	20.50									
L	9.8	81	=	10e	1e	40e	7e	9e	5e	28e
R	10.7	85	=	12e	0e	28e	10e	14e	10e	26d
Roundabouts	116.20									
A	47.7	87		20e	3d	16e	13e	12e	21e	14e
L	33.4	80		23e	4d	19d	11e	13d	16d	9d
R	35.1	85		19d	5d	18d	11e	11d	16d	19d
Bends	132.40	60	dec	24d	4d	16d	8e	24d	23e	1e
T-junctions	511.50									
A	53.1	64	=	22d	4d	19d	10e	17d	22d	5d
L	273.5	64	dec	27d	12d	18d	7e	15d	12d	10d
R	184.9	66	dec	27d	11d	17d	10e	12d	14d	9d
Crossroads	304.50									
A	129.4	62	dec	29d	0d	26d	12d	16d	16d	1d
L	91.5	64	dec	28d	11d	18d	5d	16d	12d	10d
R	83.6	71	dec	24d	10d	16d	12e	11d	17e	9d
Slowing	35.90	77	dec	53d	1e	32i	6e	4e	3e	1d
Parking	109.20	60	dec	36d	5d	17d	1e	9d	20d	13d
Reversing	70.70	80	=	26e	22e	20e	0i	8d	17i	0e

* not near junctions

Two aspects of the table are particularly striking. In the first place, before drivers become qualified to drive without supervision, some manoeuvres are performed extremely seldom. Taking the most generous definition of overtaking, so as to include, for instance, driving around all moving obstacles, pupils overtake some 50 times under supervision. Almost none of these resemble the high-speed overtaking manoeuvres typical of more experienced drivers. Similarly, drivers rarely practise lane changing. It is questionable whether such small numbers of distributed trials are sufficient to establish a consistently high level of performance. It can hardly be the case that these manoeuvres are practised under a broad range of situations, sufficient to prepare newly qualified drivers for the range of situations they will encounter later. The second finding which testifies to the need for support systems for newly qualified drivers is that, while the frequency of providing support decreases for certain manoeuvres during the course of the training programme, for others the support provided remains at a consistently high level. Even where the instructor gives a decreasing amount of support to a particular manoeuvre, certain activities within that manoeuvre may not be improving to the same extent. A case in point is support for judgment of the prevailing traffic situation. In every manoeuvre, whatever the general level of improvement, the instructor continues to consider the pupil's judgmental skills in need of support. Indeed, there are even a few activities where the perceived need for support actually increases as the course progresses.

These data indicate that, even where fully trained motorists are on the point of being licensed to drive alone, they continue to require a high level of correction and advice. It is obvious that once these motorists become fully-fledged drivers, their access to such support is minimal, if not nil.

This brief review of some of the investigations carried out as part of the GIDS project indicates that the performance of drivers differs greatly from situation to situation, and from driver to driver. It is also important to realize that, even if groups of motorists do not differ substantially in performance terms, it may be more difficult for some drivers than for others to maintain their level of performance. It is also possible that the performance level achieved is inadequate, or even dangerous, for some drivers in some circumstances. Because of this, and because of the difficulty of establishing the actual needs of different groups, we need to consider supporting individual drivers, rather than groups of motorists.

(d) Personalized Support And Learning Module (PSALM)

In the preceding sections it was concluded that support for driving performance which is tailored to the needs of groups of motorists may not be feasible and seems unlikely to promote effective learning among individual drivers. Instead, the needs of individual drivers need to be addressed, at least as an intermediate goal on the way towards a system which meets the needs of different groups of drivers. The way in which drivers should be supported depends on the behaviour that is to be supported and on the objectives the system designer has in terms of longer-term learning. The precise scheduling of support requires substantial empirical research in order to realize the full 'tutoring' potential of support systems. In order to carry out this research, a prototype support system must be available. It is essential that this support system has sufficient flexibility to allow research into the issues raised above. It is also essential that the system can be realized with existing technology, although some assumptions should be allowed regarding availability and cost reduction in future technology.

Of primary importance is a means of storing the performance profiles of individual drivers, documenting the frequency with which they have encountered particular situations and their history of 'abnormal' performance in each situation. Such a profile can easily be derived from data provided by the GIDS sensors and applications. Abnormal performance would be regarded as a deviation from a normative standard in terms of the functioning of each sensor/application, for instance, abrupt, hard braking, attempts to make fast and unusually large steering corrections, excessive acceleration and adoption of highly variable headways. It is envisaged that the data routinely collected by the GIDS sensors and other applications would be processed by a Personalized Support And Learning Module (PSALM), whether or not they have triggered a warning or other message delivered by the GIDS system. Drivers would be informed through PSALM when a criterion of abnormal performance is reached, or when they actively request support from PSALM.

Initially, estimated performance criteria, in terms of acceptable numbers of incidences of abnormal performance, might be stored along with the performance profile for each situation. Exceeding this criterion would result in a request being passed from PSALM to the Dialogue Controller, which schedules all communications to the driver for a message to be presented. Such a message would indicate that, over a period of time, there had been a tendency for (specified) deviant behaviour to occur in a particular situation, for instance, a tendency of the driver to make excessive steering corrections while taking the third exit from a four-arm roundabout. In addition PSALM would provide advice about how to correct the error.

In the short run it is expected that feedback about performance would be based on aggregated performance in a particular situation. Before such a database is built up, optimal support could not be given during a manoeuvre. Instead 'alerting messages', i.e. advance information of upcoming events, would be used 'on-line', or feedback might be provided when the driver had left the critical situation, or 'off-line' when the driver had come to a halt. As we acquire an understanding of the effects on performance of interactions between components of the GIDS system, and as knowledge grows of the appropriate time course of the messages required to support performance, it is envisaged that support for individual aspects of driver performance should become increasingly 'on-line' and adapted to that individual driver's needs. Ideally this development of the way in which the prototype can potentially aid drivers, i.e. from off-line to on-line support, will be guided by the empirical evaluations of the first GIDS prototype. It seems probable that in the future PSALM may, once a criterion is exceeded, inform the driver that performance is not adequate, and suggest a local route which requires performance of the relevant 'problem' activities. This route might then be driven for practice purposes, with PSALM giving support while the manoeuvre in question is actually being performed, so that the errors in performance were identified and corrected under PSALM supervision. These issues are to be examined in detail when the prototype PSALM system will become available as part of the DRIVE II project ARIADNE.

Conclusions and recommendations

The empirical studies discussed in the preceding sections (a) - (d) led to the following recommendations for promoting improvements in drivers' performance:

−   Driver support needs must be determined by careful consideration of the errors made by and demands placed upon drivers in specified traffic situations, rather than for the driving task as a whole;
−   Support needs of different drivers need to be established at an individual level, rather than by extrapolating from the performance of others in that situation;
−   These needs should be kept under constant review, since they are likely to change as the driver gains experience;
−   There is a need to develop support systems which 'learn' *about* the driver, thus promoting learning *by* the driver.

7.5 GIDS calibrator: individual preferences

The previous section clearly showed that the average differences between groups of drivers with certain characteristics, such as age or experience, may be smaller than the differences between members of each group. This confirms the view that, in order to make a system such as GIDS acceptable to a wide variety of drivers, it is important that each driver be given the opportunity to set GIDS functions at preferred levels. Therefore the GIDS system is equipped with a *calibrator* allowing individual drivers to adapt the system to their personal preferences. This section elaborates upon the need for and the implementation of a calibrator in GIDS. It distinguishes four categories of parameters the driver may want to change, indicates which settings should be kept within specific limits and why, and it gives guidelines concerning the calibrator interface.

The calibrator: gaining acceptance by adaptation

In-vehicle support and information systems are intended to increase traffic safety by constantly monitoring drivers and warning them when necessary. Given the variation between drivers it is evident that a single, invariable support and information system cannot provide the appropriate level of support and information to all individual drivers. For example, most young experienced male drivers may not want the system to give support at all, unless it really helps to prevent an impending collision. In contrast, elderly persons with highly limited driving experience may feel much more confident if they know that the system will give them advice and warn them of complex situations. So, on the one hand, drivers must be satisfied by a support and information system or they will not accept it. On the other hand, it is unlikely that a single, fixed-parameter system can satisfy all the needs of the broad variety of individuals in the driver population. Any developer of intelligent systems to be used by the public has to solve the problem of acceptability and, at the same time, satisfy the needs of all potential users. For GIDS the obvious solution is to adapt the system to its current user, but then the question is, how should this be done?

First, a distinction should be made between short-term and long-term adaptation (Verwey, 1990b). Short-term adaptation means that the system adapts to the current situation, that is, the momentary workload of the driver (see Chapter 4). Long-term adaptation, on the other hand, implies that the driver can make the system function as desired over a longer period, for instance an entire trip. The latter is the purpose of the GIDS calibrator. The GIDS calibrator allows drivers to install and quickly change a number of features of the driver-car interface according to their own preferences. For example, one driver may want to be addressed by a male voice and only in case of emergencies. Another driver may prefer a female voice that gives messages all the time. Another example, which seems to be more than just a matter of personal preferences, is a colour-blind driver who wants to adapt the colours used in visual messages to his specific defect in order to maximize the discriminability between signals.

Calibrator design philosophy

The basic idea is that the driver should be able to change everything in the system's functioning, except features that might give rise to dangerous situations. This possibility will increase acceptability of the system and, as a result, its use and success. The designer may constrain the freedom of the driver to some extent: safety features should not be overridden. Even so, the driver should be able to affect safety systems by being given the opportunity to change criterion values at which a warning is triggered. This possibility assumes particular importance whilst GIDS itself cannot, for the time being, adapt to characteristics of individual drivers but only to the general characteristics of groups of drivers. For instance, an older but highly skilled driver should be able to select criterion settings that are basically intended for younger drivers, if he so wishes. All the designer can do is ensure that the system warns the driver during such a manipulation that support is no longer adapted to the general capabilities of the individual's own group. Since the driver is ultimately responsible for his use of the system, this seems preferable to a situation in which the system is switched off completely due to lack of acceptance. From this point of view, the GIDS calibrator may eventually turn out to be one of the major determinants of (commercial) success for GIDS-like systems.

Four categories of settings

Basically, four categories of calibrator settings can be distinguished: those having to do with car control, features of physical messages, criteria determining when warnings are to be given, and automatic initiation of low-priority functions.

First, if GIDS is not able to determine the driver's identity and one driver may use another's smart card to calibrate system settings, it is unavoidable that someone may drive with another person's settings. Since drivers are known to use more or less automatized skills (e.g., Kelley, 1968; Verwey, 1990b), changing the basic control characteristics of the vehicle may jeopardize traffic safety (see also Section 7.3). Hence, these characteristics should not be adjustable at all. Only if it can be demonstrated that a particular feature is skill-independent should it be adaptable. When driver identification becomes possible and reliable, GIDS developers may consider permitting drivers to change settings that have to do with car control. For example, the force required to control the wheel and pedals and the kind of feedback on these controls may be selected once for a particular driver. Especially for handicapped people it may be useful to change the pedal and other control functions. Yet, with these critical settings, care should be taken that new settings are tested thoroughly on a test track before they are selected permanently.

A second category of settings refers to the physical features of messages that are not related to basic control skills: colour and intensity of the visual display, or speech characteristics of the voice generator. Since these features may affect the conspicuity of a message, the system should allow changes to these settings only to the extent that conspicuity of the message is not affected. Thus, vital messages, such as anti-collision warnings and lane-keeping warnings, should remain conspicuous while more trivial messages should not be given conspicuous properties.

Parameter values that determine if and when warning and alerting messages are to be activated should be variable between upper and lower bounds determined for all drivers. The lower bound should allow the system to warn even a highly skilled driver. The upper bound should indicate when warnings are to be given to inexperienced or older drivers. Functions in this category are, for example, anti-collision warnings and lane-keeping warnings.

Finally, some drivers may prefer to be aided automatically by some of the low-priority functions, whereas other drivers may not. Each driver should be able to set such preferences. Examples are: starting of the windscreen wipers in rain, putting on the lights at dusk, and lowering the volume of the stereo when a conversation is being held with a passenger or on the telephone, or when background noise is reduced.

Calibrator interface

A final word about the GIDS calibrator concerns the way it is operated. Since setting or resetting the system will happen only occasionally, and since it is not desirable that drivers pay attention to the calibrator while driving or use it as a toy during longer journeys, it is recommended that the calibrator be usable only when it is entirely safe to do so, for instance, when the car is parked or being driven on a test track.

In addition, the calibrator should be simple and convenient to operate by experienced as well as inexperienced GIDS users. Therefore, it must make use of a menu-driven interface with an on-line help function. Control of the menu should be accomplished with only a few keys on the keyboard. The calibrator should indicate clearly what settings have been changed before they are made permanent.

Conclusions and recommendations

The following recommendations for the GIDS calibrator can be derived from these considerations:

- Intelligent systems to be used by the public should be both acceptable and satisfactory for all potential users. For the GIDS system the obvious solution is to make the system adaptable to its current user by means of a calibrator;
- The GIDS calibrator allows drivers to select and quickly change a number of features of the driver-car interface according to their own preferences. The driver should be able to change everything in the system's functioning, except features that may give rise to or sustain dangerous situations;
- If drivers install a critical setting which is not meant for the group of drivers to which they belong they should be informed, testing on a test track may be required;
- The calibrator should be set or modified only when it is entirely safe to do so, for instance, while the car is parked or being driven on a test track.

7.6 Integrating functions in the GIDS system

The preceding sections have discussed a variety of guidelines for individual GIDS functions. However, it will be clear that the separate functions should not only be independently adapted to human capabilities, they must also be integrated into one transparent and easy-to-use system. Hence, when designing a complex system such as GIDS it is essential that potential bottlenecks in the information flow between the driver and the conceptualized GIDS system are identified. On the basis of a questionnaire distributed amongst the GIDS partners, Verwey (1990a) proposed an integrated interface design in which modalities and priorities were assigned to different functions in the system. The results of that discussion are summarized in Table 7.3. The table shows each function with a number of so-called interaction clusters (see Chapter 4). An interaction cluster can be defined as a more or less fixed sequence of messages and driver actions used in interaction with one function of the system, that is, a subsystem or application. The basic idea behind the table is that interaction clusters for which there is no explicit need during actual driving, should be available only when the vehicle is standing still (cluster 1a, 1c, 1d, 3a, 4a, 4c, 5a in Table 7.3). Some interaction clusters should be possible when driving, e.g., when they rely on speech commands (1a, 1c). The remaining interaction clusters should be amenable to dynamic scheduling in the GIDS system (1b, 2a, 3b, 3c, 6, 7a, 7b, 8a, 8b, 8c).

In this arrangement, the number of potential bottlenecks is highly limited. As shown in Table 7.3, the only visual information presented by the GIDS system during driving is route guidance and speed and headway support. Visual messages, therefore, are not likely to interfere much with each other. Auditory messages may be given by the route guidance system and by some low-priority applications (radio, telephone). Therefore, the GIDS system has important functions in presenting visual information only when visual demands of driving are low, and in keeping auditory messages separated. Finally, the driver may be warned by an increasing counterforce on the accelerator pedal of an approaching object on a collision course when headway is too short, or when speed is inappropriate. Improper lane keeping is indicated by counterforce on the steering wheel. Given the likely increase in workload when messages are presented in a different format or modality than they normally are (Verwey, 1990a), this possibility should only be implemented when further research has indicated that the rise in workload is limited. For the present it is not recommended for the GIDS system.

Table 7.3 Likely driver interaction clusters as a function of input/output modality and priority

Interaction cluster	Highest priority allowed	System to driver	Driver to system
1. Route guidance			
a. entering destination	*	V	Mk/S
b. on-line guidance route information	3	A/V	
c. asking information	*		Mt/S
d. information presentation	*	V	
2. Collision avoidance			
a. warning the driver	6	Tg	
3. Control support			
a. select lane-keeping support	*	V	Mk
b. lane-keeping support	5	Tw	
c. speed and headway support	6	Tg/V	
4. Performance evaluator			
a. selecting	*	V	Mt
b. scanning driver performance			
c. presenting feedback info	*	A/V	Mt
5. GIDS calibrator			
a. setting parameters/preferences	*	V	Mk
6. Repeat last message	depends	A/V	Ms/S
7. Telephone			
a. dialling	1	Ts	Ms/S
b. conversing	2	A	S
8. Stereo			
a. tuning	1	A	Ms/S
b. volume control	2	A	Ms/S
c. changing cassette	1	T/A	M

An asterisk (*) indicates driver-car interaction while stationary. Lower numbers indicate lower priority levels (see Table 4.1). Driver input involves auditory input (A), visual input (V), and tactile input on the wheel (Tw), the accelerator pedal (Tg) and switches/buttons (Ts). Driver output incorporates mainly manual (M) output, but pedal (P) and speech (S) output may also be included. Manual driver output can be subdivided into keyboard (Mk), switches/buttons (Ms) and touch screen (Mt).

Car control is primarily manual. This may produce (motor) interference between normal car control and the manual control of GIDS functions. Therefore, all GIDS functions to be used during driving should also be controllable by speech. This would protect the driver from being forced to control several GIDS functions by hand simultaneously. When most functions can be controlled both by speech and by hand, the individual driver may learn to exercise control over several systems concurrently by using different control modalities. In this case the bottleneck in controlling the car and the GIDS functions will lie mainly in drivers' cognitive resources, rather than in their motor resources (see Chapter 3), and the GIDS system should only suppress messages that require cognitive resources for subsequent control actions while these are also being used for other vehicle or GIDS control tasks.

The protocol shown in Table 7.3 represents a rather conservative picture of the adaptibility of the GIDS system. For example, it does not allow the driver to manually enter a destination in the route guidance application while driving. Research should indicate the extent to which a driver should be allowed to perform more functions while driving. For example, in future versions of GIDS a destination might possibly be entered *en route* if the driver is able to do this by simply typing the destination address or postal code, but not if a map has to be consulted. In Chapter 8 the implementation of these integrative guidelines in the GIDS system will be described.

Conclusions and recommendations

These notions of integrating several tasks give rise to the following recommendations:
—   If several applications may present information at the same time, care should be taken that the messages are scheduled according to their priority with respect to safety;
—   In the GIDS system optimal use should be made of the possibility to present information in modalities other than the visual one;
—   Interaction clusters (i.e., sequences of messages and actions pertaining to one application) should be concluded before messages from another application with a similar or lower priority are presented;
—   Messages should not be unexpectedly presented in another format or modality until research has indicated that workload will not increase as a result;
—   To increase the possibility of concurrent control of several applications they should be controllable both manually and by speech.

References

Alm, H. (1990). Drivers' cognitive models of routes. In W. van Winsum, H. Alm, J.M.C. Schraagen, & J.A. Rothengatter (Eds.), *Laboratory and field studies on route representation and drivers' cognitive models of routes* (pp. 35-48). Deliverable Report DRIVE V1041 GIDS/NAV2. Haren, The Netherlands: Traffic Research Centre, University of Groningen.
Alm, H., & Berlin, M. (1991). Optimal amount of verbal navigation information. In J.M.C. Schraagen, H.Alm, M. Berlin, J.A. Rothengatter, & P. Westerdijk (Eds.), *Effects of navigation information aspects.* (pp. 15-22). Deliverable Report DRIVE V1041 GIDS/NAV 04. Haren, The Netherlands: Traffic Research Centre, University of Groningen.

Davis, J.R., & Schmandt, C.M. (1989). The back seat driver: Real time spoken driving instructions. In D.H.M. Reekie, E.R. Case, & J. Tsai (Eds.), *Proceedings of the first Vehicle Navigation and Information Systems Conference (VNIS'89)* (pp. 146-150). Toronto: IEEE.

Evans, L., & Wasielewski, P. (1983). Risky driving related to driver and vehicle characteristics. *Accident Analysis and Prevention, 15,* 121-136.

Färber, B., Färber, Br.A., Godthelp, J., & Schumann, J. (1990). *State of the art and recommendations for characteristics of speed and steering support systems.* Deliverable Report DRIVE V1041 GIDS/CON1. Haren, The Netherlands: Traffic Research Centre, University of Groningen.

Färber, B., Godthelp, J., Schumann, J., & Wontorra, H. (1991). *Predictions of effects of prototype implementation on handling aspects of driving tasks.* Deliverable Report DRIVE V1041 GIDS/CON2. Haren, The Netherlands: Traffic Research Centre, University of Groningen.

Färber, B., Naab, K., & Schumann, J. (1991). *Evaluation of prototype implementation in terms of handling aspects of driving tasks.* Deliverable Report DRIVE V1041 GIDS/CON3. Haren, The Netherlands: Traffic Research Centre, University of Groningen.

Fenton, R.E. (1966). An improved man-machine interface for the driver-vehicle system. *IEEE Transactions on Human Factors in Electronics, 7,* 150-157.

Godthelp, J. (1990). *The use of an active gas-pedal as an element of an intelligent driver support system; literature review and explorative study.* Report IZF 1990 B-17. Soesterberg, The Netherlands: TNO Institute for Perception.

Godthelp, J., & Schumann, J. (1991). The use of an intelligent accelerator as an element of a driver support system. *Proceedings 24th ISATA International Symposium on Automotive Technology and Automation* (pp. 615-622). Croydon: Automotive Automation Ltd.

Groeger, J.A. (1991). Supporting training drivers and the prospects for later learning. Commission of the European Communities, Directorate-General Telecommunications, Information industries and Innovation (Ed.), *Advanced telematics in road transport* (Vol. 1, pp. 314-330). Amsterdam: Elsevier.

Groeger, J.A., & Grande, G.E. (1991) Support received during drivers' training. In M.J. Kuiken & J.A. Groeger (Eds.), *Report on feedback requirements and performance differences.* Deliverable Report DRIVE V1041 GIDS/ADA2. Haren, The Netherlands: Traffic Research Centre, University of Groningen.

Groeger, J.A., Kuiken, M.J., Grande, G.E., Miltenburg, P.G.M., Brown, I.D., & Rothengatter, J.A. (1990). *Preliminary design specifications for appropriate feedback provision to drivers with differing levels of traffic experience.* Deliverable Report DRIVE V1041 GIDS/ADA1. Haren, The Netherlands: Traffic Research Centre, University of Groningen.

Hess, R.A., & Modjtahedzadeh, A. (1990). A control theoretic model of driver steering behaviour. *IEEE Control Systems Magazine, 10,* 3-8.

Hosman, R.J.A.W., & Van der Vaart, J.C. (1988). Active and passive side stick controllers: Tracking task performance and pilot control behaviour. *Proceedings AGARD Conference on Man-machine Interface in Tactical Aircraft Design and Combat Automation.* AGARD-C8-424, 26-1/26-11.

Jagacinski, R.J., Flach, J.M., & Gilson, R.D. (1983). A comparison of visual and kinaesthetic-tactual displays for compensatory tracking. *IEEE Transactions on Systems, Man, and Cybernetics, 13,* 1103-1112.

Janssen, W.H., & Nilsson, L. (1990). *An experimental evaluation of in-vehicle collision avoidance systems.* Deliverable Report DRIVE V1041 GIDS/MAN2. Haren, The Netherlands: Traffic Research Centre, University of Groningen.

Kelley, C.R. (1968). *Manual and automatic control.* New York: Wiley.

Koning, G.J., & Bovy, P.H.L. (1980). *Routeanalyse, een vergelijking van model en enquêteroutes van autoritten in een stedelijk wegennet.* (Memorandum 24). Delft, The Netherlands: Delft University of Technology.

Kuiken, M.J., Miltenburg, P.G.M., de Waard, D., & van der Mey, P. (1991). Behaviour of novice, inexperienced and experienced drivers during manoeuvring tasks. In M.J. Kuiken & J.A. Groeger (Eds.), *Report on feedback requirements and performance differences.* Deliverable Report DRIVE V1041 GIDS/ADA2. Haren, The Netherlands: Traffic Research Centre, University of Groningen.

Nilsson, L., Alm, H., & Janssen, W.H. (1991). *Collision avoidance systems: Effects of different levels of task allocation on driver behaviour.* Deliverable Report DRIVE V1041 GIDS/MAN3. Haren, The Netherlands: Traffic Research Centre, University of Groningen.

Panik, F. (1984). Fahrzeugkybernetik. *Proceedings XX FISITA-congress, Vienna, 1984.* Society of Automotive Engineers Special Publication SAE-P-143, 0, 3.273-3.284.

Rockwell, T.H. (1972). Skills, judgement and information acquisition in driving. In T.W. Forbes (Ed.), *Human Factors in highway traffic safety research* (pp. 133-164). New York: Wiley-Interscience.

Rühmann, H. (1981). Schnittstellen in Mensch-Maschine-Systemen. In H.Schmidtke (Ed.), *Lehrbuch der Ergonomie* (2nd ed.) (pp. 351-376). München, Germany: Hanser.

Rule, R.G., & Fenton, R.E. (1972). On the effects of state information on driver-vehicle performance in car following. *IEEE Transactions on Systems, Man, and Cybernetics, 2,* 630-637.

Sanders, M.S., & McCormick, E.J. (1987). *Human factors in engineering and design* (6th ed.). New York: McGraw-Hill.

Schraagen, J.M.C. (1990). *Strategy differences in map information use for route following in unfamiliar cities: Implications for in-car navigation systems.* Report IZF 1990 B-6. Soesterberg, The Netherlands: TNO Institute for Perception.

Schraagen, J.M.C. (1991). *An experimental comparison between different types of in-car navigation information.* Report IZF 1991 B-1. Soesterberg, The Netherlands: TNO Institute for Perception.

Schumann, J., Färber, B., & Wontorra, H. (1991, March). *Das Lenkrad als propriozeptiv-taktiles Display.* Paper presented at the 33th 'Tagung experimentell arbeitender Psychologen', Giessen, Germany.

Schumann, J., Godthelp, J., Färber, B., & Wontorra, H. (1993). Breaking up of open-loop steering control actions: The steering wheel as an active control device. *Vision in Vehicles IV.*

Sheridan, T.B., & Ferrell, W.R. (1974). *Man-machine systems: Information, control, and decision models of human performance.* Cambridge, MA: MIT Press.

Streeter, L.A., Vitello, D., & Wonsiewicz, S.A. (1985). How to tell people where to go: Comparing navigational aids. *International Journal of Man-Machine Studies, 22,* 549-562.

Van Winsum, W. (1989). Route choice criteria for car drivers: A review of the literature. In J.A. Rothengatter (Ed.), *Navigation information requirements: A literature review.* Deliverable Report DRIVE V1041 GIDS/NAV1. Haren, The Netherlands: Traffic Research Centre, University of Groningen.

Van Winsum, W. (1990). A MAUT study on car drivers' preferences of routes. In W. van Winsum, H. Alm, J.M.C. Schraagen, & J.A. Rothengatter (Eds.), *Laboratory and field studies on route representation and drivers' cognitive models of routes.* Deliverable Report DRIVE V1041 GIDS/NAV2 (pp. 11-33). Haren, The Netherlands: Traffic Research Centre, University of Groningen.

Verwey, W.B. (1989). Simple in-car route guidance from another perspective: Modality versus coding. In D.H.M. Reekie, E.R. Case, & J. Tsai (Eds.), *Proceedings of the first Vehicle Navigation and Information Systems Conference (VNIS'89)* (pp. 56-60). Toronto: IEEE.

Verwey, W.B. (1990a). *Adaptable driver-car interaction in the GIDS system: Guidelines and a preliminary design from a human factors point of view.* Report IZF 1990 B-12. Soesterberg, The Netherlands: TNO Institute for Perception.

Verwey, W.B. (1990b). *Adaptable driver-car interfacing and mental workload: A review of the literature.* Deliverable Report DRIVE V1041 GIDS/DIA1. Haren, The Netherlands: Traffic Research Centre, University of Groningen.

Verwey, W.B. (1991). *Towards guidelines for in-car information management: Driver workload in specific driving situations.* Report IZF 1991 C-13. Soesterberg, The Netherlands: TNO Institute for Perception.

Verwey, W.B. (1992). *Further evidence for benefits of verbal route guidance instructions over spatial symbolic guidance instructions.* Report IZF 1992 C-4. Soesterberg, The Netherlands: TNO Institute for Perception.

Verwey, W.B., & Janssen, W.H. (1989). Driving behaviour with electronic in-car navigation aids. *Proceedings of the INRETS/VTI Congress Road Safety in Europe.* (Report 342A, pp. 81-97) Gothenburg, Sweden: VTI.

Wickens, C.D. (1984). *Engineering psychology and human performance.* Columbus, OH: Merrill.

Chapter 8
GIDS architecture

Ep H. Piersma, Sjouke Burry, Willem B. Verwey, Wim van Winsum

8.0 Chapter outline

Chapter 7 has dealt with the functions of the GIDS prototype. The present chapter will describe its actual implementation at the software and hardware levels. The software modules and their functions will be introduced in Section 8.1 and subsequently described in more detail in Sections 8.2 to 8.6. In Section 8.2 the connection to the outside world is covered. The software dealing with the driving task is described in Section 8.3. Section 8.4 is concerned with PSALM, the Personalized Support and Learning Module. Section 8.5 describes the user interface of the GIDS system and the way in which workload adaptivity is implemented. Section 8.6 introduces the active controls. This will be followed in Section 8.7 by identifying the allocation of the various software modules in the GIDS system to various pieces of hardware used in both versions of the prototype. One is embedded in the Traffic Research Centre (TRC) simulator, a fixed-base driving simulator, including a dynamic simulation of the Small World with autonomous traffic (see Chapter 9), and one embedded in ICACAD (Instrumented Car for Computer Assisted Driving), the test vehicle of TNO Institute for Perception. These prototypes differ only in terms of the way navigation and sensory functions are implemented, as described below. Finally Sections 8.8 and 8.9 describe the specifics of the two respective GIDS system implementations.

8.1 Introduction

The previous chapter has described the many functions to be found in the GIDS system and their empirical foundation. The question to be answered in the present chapter is what architecture can be devised, given the current state of the art of software and hardware engineering, that can exhibit those functions. We will describe the form of such an architecture, discuss its several components, and present the first implementations of the architecture in the GIDS prototypes.

Generic design

It is important to note that, in a sense, the physical realization of the GIDS prototype is arbitrary. For example, the number and type of processors which run the various procedures that constitute the GIDS intelligence is immaterial (at least for the purposes of research and development) as long as the specified functions are achieved. The particular choices to be discussed have been made largely for pragmatic reasons: ease and speed of implementation. In principle there are numerous other ways of achieving the appropriate set of functions. In other words, the design is generic.

There are now two realizations of the GIDS principle, two identical GIDS systems, but they are connected in different ways to different worlds. One is embedded in the Traffic Research Centre's driving simulator, connected to a simulated dynamic interactive traffic world. This is known as the Small World Simulation and it is described in detail in Chapter 9, the other forms part of the ICACAD experimental vehicle developed by the TNO Institute for Perception for use in real-world situations.

Conceptualizing GIDS

The early conception of the GIDS system involved a set of loosely coupled 'applications' that fed their high-level results to the central GIDS intelligence for filtering and processing. Thus, for instance, a radar-based collision avoidance system might send the message "Slow lead vehicle ahead, decelerate to 10 km/h within 3 seconds." to GIDS for presentation to the driver. The trouble encountered with this approach resides in the fact that many separate functions overlap in their processing requirements; consider, for instance, collision avoidance and headway support. A seamless integration of such functions, especially those concerning speed and heading support, requires an integrated determination of the acceptable ranges of speed and direction (Piersma, 1990). For example, if two subsystems independently evaluate speed with respect to the traffic situation, taking into account some but not all relevant traffic objects on the basis of possibly different criteria (e.g., a worst case versus some acceptable time-to-collision criterion), then a module integrating alternative (and possibly contradictory) support messages will have to consider the actual reasoning that underlies these messages in order to evaluate the traffic situation. Therefore it is more appropriate to have a centralized determination of acceptable speed and course, taking into account all available evidence regarding the traffic situation.

Another shift in thinking occurred as the specification matured. An initial vague notion of a 'monolithic' know-it-all intelligence, capable of reasoning about every detailed aspect available in the data of any application or in a generic database of facts about the world, was deemed non-implementable because of an exponential increase in the possible combinations of data to be processed. This know-it-all notion has been replaced by a notion of strict modularization, with individual modules each exhibiting a limited and unique intelligence. The proper cooperation between these modules should provide the system as a whole with the proper level of intelligence.

Such a modularization of the GIDS system still leaves many degrees of freedom when it comes to actual functions. In fact, the detailed specification of the support behaviour of the system far exceeds the present level of modularization. The current architectural design, however, may well survive many changes in functional requirements, with respect to the I/O devices used, type of support given, and degree of adaptation to individual needs.

The architecture in outline

We will now present an overview of the various modules of the GIDS system (Piersma, 1990). First, a modular design requires a communication protocol. The conceptual model for communication between the modules adopted in GIDS is message-passing between processes. Each module is able to exchange messages with all of the other modules simply by generating the message in the appropriate format with the appropriate address and posting it.

Figure 8.1 presents the GIDS prototype system. A navigation system and any number of sensors provide the driving model in the Manoeuvring and Control Support Module with information about the road segments in the vicinity of the GIDS car, data about relevant static and dynamic traffic objects, and data about the status of the car controls. The Manoeuvring and Control Support Module (MCSM) determines which support messages are required. These messages will usually be presented through the active controls (accelerator pedal and steering wheel), but possibly also by means of spoken messages. The required messages are passed on to the Scheduler so that decisions can be made about when they will actually be presented. In addition the MCSM determines preliminary timings for navigation messages provided by the Navigation module. These 'timings' consist of the interval within which the message must be presented plus a preferred moment within that interval. The Scheduler will decide whether the message can be presented at the preferred time or must be scheduled elsewhere in the specified interval. Finally the MCSM identifies the tasks the system knows the driver is engaged in, or anticipates the driver will be engaged in shortly. This information is the input to the

Workload Estimator. The Workload Estimator, in turn, provides the Scheduler with workload constraints that allow the latter to present or suppress messages in order to keep mental workload within acceptable limits. The Scheduler receives messages to be presented to the driver from the MCSM and the Dialogue Generator, along with expected workload, preferred presentation time, relevance interval and message importance. The Scheduler advances, postpones, or suppresses messages as workload requires. The Scheduler commands the Dialogue Generator or Car Body Interface (in the case of the active controls) to present messages to the driver. The Dialogue Generator, finally, consists of the user interface and the secondary applications (tuner and telephone).

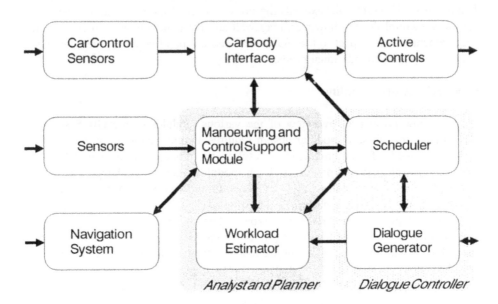

Figure 8.1 The GIDS prototype system (see text for explanation)

The functions of the various modules have been determined by a clear and consistent design philosophy: restricting the types of data processed by a module reduces the need for communication (not all data need be sent to each module) and it also reduces the necessary processing (fewer types of data implies less diversity in processing). In more pretentious terms: each module contains a unique part of the intelligence of the system as a whole, allowing implementation within the necessary limits on complexity. Thus, the processing in the navigation system concerns only the current position in the road network and the route to follow; the processing in the Analyst/Planner concerns only the immediate future with respect to the driving task; the processing in the tutoring module (PSALM) concerns only trends in driving performance and appropriate instruction; the processing in the Dialogue Controller concerns two main functions: one, the Scheduler, is concerned only with scheduling messages, taking into account relative importance and workload; the other function, the Dialogue Generator, generates dialogues for the control of specific functions.

8.2 Connection to the outside world: sensors and navigation system

The sensors in the GIDS prototype are of two types. One senses the actual use of car controls by the driver. The other senses the relevant objects in the traffic environment and some of their interrelationships. In addition the navigation system computes the route to follow to the destination. On the road it keeps track of the position of the car in the road network. The road network is represented by a set of segments that represent parts of the network, such as a stretch, a curve, or an intersection. The navigation system informs the MCSM of the current and adjacent segments, indicating the road type and condition of each segment and of the route to follow.

In the GIDS project there have been two parallel attempts at developing appropriate sensors for vehicle and obstacle detection. The first, by Philips, is based on radar technology, whilst the other, by Renault, applies laser technology. Altogether, however, it was found to be impractical to incorporate either system into a prototype in the short run. In the current ARIADNE project, in which the GIDS concept is being developed further, a more explicit effort is being made at supplying a radar-based object detection and tracing system by Philips Research Laboratories UK, and British Aerospace Sowerby Research Centre, UK.

Since it is not crucial for the GIDS project to specify in detail how, technically speaking, the data to be processed are produced, as long as a minimal informational basis is provided for, it was considered appropriate to use simulated sensors in the ICACAD implementation (see Section 8.9). The Small World simulation, on the other hand, is capable of providing any (simulated) sensory information that is useful. This allows a detailed demonstration of the GIDS system's behaviour in a context in which an informationally rich set of sensors is available.

A similar story applies to the navigation system. Several navigation systems are commercially available today. Therefore developing yet another such system would have been a complete waste of time. Implementing an interface to one of the existing systems, with all the associated difficulties of gaining access to confidential specifications of those systems, would have been a major task. Although technically speaking it would have been completely feasible, for the time being only a simple navigation system encompassing the experimental and demonstration routes (see Chapters 9 and 10) has been incorporated in the GIDS prototype.

Sensors

The sensors pick up sensory information and transmit this information to the GIDS system, once every 0.1 seconds. This information differs between the two proto-type systems, as indicated below.

TRC-Simulator

In the TRC-Simulator version of the prototype the following information is sent to the Manoeuvring and Control Support Module:

Car control sensor data
–   steering wheel angle, in radians (with respect to heading of car);
–   brake, percentage pressed down;
–   clutch, percentage pressed down;
–   accelerator pedal, percentage pressed down;
–   indicator switch (Off, Left, Right).

Car sensor data
–   road segment identification of the segment in which the car is currently travel-ling. A segment is a part of a road, between two intersections or other segments, that is either straight or curved;
–   speed of the car, in cm s^{-1};
–   acceleration of the car, in cm s^{-2};
–   heading of the car with respect to the tangent of the road, in radians;
–   distance to the end of the segment from the front of the car, in cm, along the path of the road, where the distance is normalized over the centre of the right lane;
–   lane the car is currently in (Right shoulder, Right lane, Left lane, Left shoul-der);
–   lateral distance, measured perpendicular to the direction of the road, from the centre of the front of the car to the right side of the right lane, in cm;
–   lateral velocity of the car with respect to the right side of the right lane, in cm s^{-1}.

Dynamic data
–   traffic light indication for first traffic light on path ahead (No traffic light, Green, Amber to red, Red, Flashing amber, Black).

Other traffic in vicinity of the car
—   number of vehicles (within range of sensor).

For each simulated vehicle the following variables are specified:
- relative position (Lead, Rear, Oncoming, From left, From right, From ahead);
- speed, in cm s^{-1};
- acceleration, in cm s^{-2};
- distance, in cm; includes distance from front bumper to rear bumper of lead vehicle, distance from rear bumper to front bumper of rear vehicle, from front bumper to front bumper of oncoming vehicle, and, for vehicles from left, right, or ahead: distance to intersection;
- lane the car is currently in (Right shoulder, Right lane, Left lane, Left shoulder);
- lateral distance, measured perpendicular to the direction of the road, from the centre of the front of the car to the right side of the right lane, in cm;
- lateral velocity of the car with respect to the right side of the right lane, in cm s^{-1};
- indicator (Off, Left, Right);
- vehicle length, in cm;
- vehicle width, in cm.

ICACAD
In the ICACAD version of the prototype the following information is sent to the Manoeuvring and Control Support Module by the ICACAD instrumentation computer (described in Section 8.9):
- road segment identification of the segment in which the car is currently travelling;
- speed of the car, in cm s^{-1};
- acceleration of the car, in cm s^{-2};
- distance from the start of the segment, in cm;
- indicators (None, Right blinkers, Left blinkers, Lights off, Lights dimmed, Lights on);
- lead vehicle detection (Detection off, On and lead detected, On and no lead detected);
- distance to lead vehicle, in cm, if lead vehicle detected;
- lead vehicle speed, in cm s^{-1}, if lead vehicle detected.

Navigation

The MCSM requests information from the Navigation module concerning road layouts in the vicinity of the GIDS car. In the TRC-simulator, the simulator provides this function, in the ICACAD version the Navigation Module proper provides information for the fixed experimental route, as specified in a datafile.

The GIDS system can request either road segments information or intersection information that contains the following data.

Information on road segments
Information is returned on all sequential segments from the current (or indicated) segment to the next intersection on the path the GIDS car is to follow.

The data include the number of segments in data package plus for each segment in the package:
−   segment identification;
−   segment type (stretch or curve);
−   segment width, in cm;
−   segment length, in cm;
−   radius of segment, in cm, if segment is curved;
−   turn, whether the segment turns left or right if the segment is curved;
−   turn angle, the angle difference between start - and end angle of curved segment;
−   the next segment identification; if it is the last segment in the list, then the intersection identification is given;
−   a flag indicating whether overtaking is permitted;
−   the minimum speed allowed, in cm s^{-1};
−   the maximum speed allowed, in cm s^{-1}.

Information on intersections:
Information is sent on all possible paths through the intersection from the perspective of the GIDS car's current direction of travel. These tracks are listed, counterclockwise from the path the car is currently on, in the order Right, Straight, Left. The data include:
−   intersection identification;
−   type of intersection (Roundabout, T-junction, Crossroads).

plus for every track:
−   width, in cm;
−   length, in cm;
−   radius in, cm;
−   turn angle, the angle difference between start- and end angle of curved segment;
−   turn, whether the segment turns left or right if the segment is curved;
−   indication of right of way (depending on traffic signs);
−   whether entry is allowed in the path where the track leads;

- direction of the track (Turn-right, Straight-ahead, Turn-left);
- segment identification of the first segment after the track;
- maximum speed allowed on the intersection, in cm s^{-1};
- minimum speed allowed on the intersection, in cm s^{-1};
- whether overtaking is permitted on the intersection;
- an indication whether the intersection is controlled by traffic lights;
- the required direction of the GIDS car (Straight-ahead, Turn-left, Turn-right);
- the exit number if the intersection is a roundabout.

8.3 Supporting the driving task: Analyst/Planner

The Analyst/Planner consists of two modules, the Manoeuvring and Control Support Module (MCSM) and the Workload Estimator.

The Manoeuvring and Control Support Module oversees the immediate future with respect to the driving task. From the available sensory and navigational information it computes whether any rules for acceptable driving have been violated, and require a support message (see Chapter 6). Second, it times the navigation messages, that is, it determines the relevance and preferred timing to be passed on to the Scheduler (see below). Finally it anticipates driver actions for the sake of the Workload Estimator (see Section 8.5 for details).

The Workload Estimator is the module within the Analyst/Planner that estimates the workload for the next few time-frames, based on the current traffic situation and actual and anticipated driver actions and the types of road segments to be encountered next. It informs the Scheduler accordingly. The Workload Estimator is further explained in Section 8.5 along with the Scheduler.

8.4 Supporting and improving the driver qua driver: PSALM

PSALM, the Personalized Support and Learning Module, (Groeger et al., 1990; see also Chapter 7) was designed to detect trends in driving performance, based on data provided by the MCSM and, when appropriate, to send a request to the Dialogue Controller for presenting a message, explaining "the right way to drive" with respect to the error made. This design is shown in Figure 8.2. PSALM should also provide after-trip statistics and comments about driving performance.

The presently available function, however, provides the driver with anticipatory spoken warnings regarding the upcoming traffic situation, for example: "Round about ahead", "Obstacle ahead". Such messages are not in any sense contingent on the performance of the driver on the tasks concerned and are only optionally available, intended for use by novice drivers.

For pragmatic reasons, rather than any inherent impracticality in the proposed function, personalized instructional support was not implemented in the GIDS prototype. This would have allowed support for performance that is adapted to each individual's previous history of success or failure on a particular manoeuvre. The implementation of PSALM is part of the workplan for the GIDS follow-up project, ARIADNE, in which the GIDS system prototype Mark II will be developed.

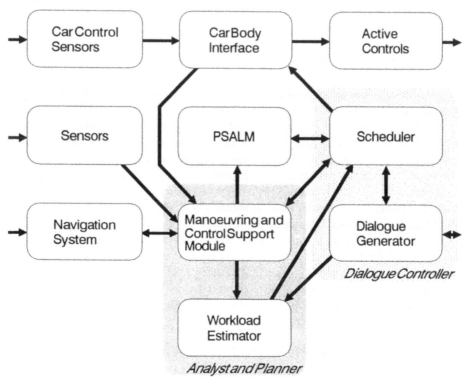

Figure 8.2 The design blueprint for the GIDS prototype system. It differs from Figure 8.1 only by the presence of the Personalized Support and Learning Module (PSALM)

8.5 The Dialogue Controller

The requests for presentation of a message to the Dialogue Controller contain an estimate of the workload the message will impose and an indication of the importance of the message. The Dialogue Controller schedules messages using the workload estimates from the Analyst/Planner, keeping workload within bounds by advancing, postponing, or even suppressing messages. A second function of the Dialogue Controller is to control the I/O devices and applications. Last but not least the Dialogue Controller generates the dialogues that allow the driver to interact with the various functions that are integrated in the system (e.g., telephone, navigation system, radio). The Dialogue Controller consists of two modules: the Dialogue Generator and the Scheduler.

Dialogue Controller

The Dialogue Controller provides a first-order approximation to workload adaptivity (Piersma, 1991; see also Chapter 7). The idea of workload adaptivity is twofold. First, it deals with the degree to which a task currently performed allows the agent to be engaged, simultaneously, in other activities. In principle, it is possible to estimate with some accuracy the so-called workload at every instant in task performance. Thus workload is seen as the complement of 'spare capacity'. Second, the presentation of information to a driver can be thought of as an additional task, since the driver has to perceive, process, and perhaps act upon the information presented. When the driver's workload is high, we would expect either driving (primary task) or processing the support message (secondary task), or both, to deteriorate. From this point of view, workload adaptivity for the GIDS system essentially consists of organizing messages to be presented when workload allows them to be fully processed, or even suppressing messages until some time when workload is low enough again. In the current GIDS prototype, workload is modelled in a simple way, by treating the driver as a set of modalities or information-processing resources, each of which can be loaded separately (e.g., cognitive, visual, auditory, tactile). In other words, the driver is assumed to possess a series of independent resources (Chapters 3 and 4).

The workload demands of a range of Small World driving tasks have been measured (Verwey, 1991) and average estimates derived from these measurements converted into workload constraints. This procedure determines what types of messages (in terms of modality and complexity) are allowed in each situation (see Chapter 7).

The ringing of a car telephone during the approach to an intersection provides a simple example of the way in which workload adaptivity works. Whilst driving towards an intersection the driver is visually and cognitively loaded as a result of his scanning the environment and deciding if and when to yield to other traffic. The sound of a ringing telephone would induce an extra cognitive load, namely that of deciding whether to answer the call and have a conversation. This could increase the driver's workload beyond acceptable levels. Therefore, the ringing of the telephone is postponed until just after the intersection, assuming that the workload caused by the driving task is down to an appropriate level.

Two remarks are in order. First, each type of message has been assigned channels through which it will always be communicated, given certain user preferences.

Smiley and Michon (1989) have suggested dynamically determining the channels through which a message is communicated to the driver. However, since the degree of familiarity with the type and content of a message is an important determinant of the workload it induces (Verwey, 1990), this suggestion has not been adopted for the GIDS prototype. Second, a careful design of support through multiple modalities will partially solve the workload problems associated with the presentation of multiple messages in close sequence. This is because messages can be made to interfere less frequently at the sensory level. For example, feedback on the accelerator should interfere only minimally with a spoken navigation message, as compared, for instance, with a beep or a buzz.

Scheduler
The Scheduler is a module in the Dialogue Controller that accepts requests (for presentation of a message) and schedules the messages within the constraints imposed by the workload. Each request comes with workload estimates and importance indication attached. All messages that are, in principle, subject to advancement, postponement, or suppression are processed by the Scheduler. There are multiple sources for requests of various kinds. In the first place there are navigation instructions as produced by the MCSM on the basis of information from the navigation system. Second, there are warnings from the MCSM. Provisions have also been made to accept instructions from PSALM. Finally there are the dialogue outputs from the Dialogue Generator, for instance, feedback on recognized commands spoken by the driver.

Workload adaptivity is achieved in the Scheduler by suppressing support messages or timing them for presentation, on the basis of the workload constraints determined by the Workload Estimator. A message to be presented to the driver is passed on to the Scheduler to be timed, along with its relevance interval, preferred time of presentation, importance, and associated expected workload per modality. The relevance interval specifies the interval outside of which the message should never be presented, for lack of relevance. The preferred time of presentation is the timing one would choose for a certain type of message to obtain maximal consistency in the timing of presentations, disregarding workload and possibly occurring other messages. The Workload Estimator is basing the workload constraints on actions anticipated by the MCSM. In the GIDS prototype, the Workload Estimator is informed of the following actions:
–   Following the road;
–   Turning;
–   Negotiating intersection;
–   Traversing intersection;
–   Negotiating curve;
–   Stopping;
–   Slowing down;

- Moving off;
- Lane changing;
- Following a lead vehicle;
- Presence of a tailgater;
- Being overtaken;
- Overtaking;
- Waiting at traffic lights;
- Waiting for a lead vehicle to move;
- Waiting for a gap;
- Avoiding an obstacle;
- Engaging in a phone conversation.

One or more of these actions may be current at a particular point in time, or in the time window processed by the Workload Estimator. A table relates experience (trainee, novice, experienced, very-experienced) and 'action' to workload constraints. If at any particular moment more than one action may be performed, the values of the most constraining action are used. Workload constraints may exist for each information processing resource (cognitive, auditory, visual, voice, haptic feet, haptic hands, output hands, output feet). The first three resources are of great importance, since they actually constrain the length and number of messages allowed, whilst the latter resources function only to prevent multiple messages occurring at once. Constraints are specified for each resource by constraint pairs. A constraint pair for a particular resource at a particular instant specifies: (1) the maximum time allowed for a load on this resource by a support message starting at that instant; (2) the minimum time before the resource may be loaded with subsequent support messages if the load on the resource by a support message ends that instant. The latter should allow the user to do something else besides processing support messages, by avoiding uninterrupted presentation of successive messages. When a message is to be presented at some point ahead in time, the message is timed for presentation in accordance with prevailing workload constraints. Exceptions to this are made when no constraint satisfaction is possible and the message priority is very high as, for instance, in an acutely hazardous situation.

Dialogue Generator

The Dialogue Generator communicates with the driver and with the interactive GIDS modules, such as the Navigation system (e.g., for entering a destination) or the telephone (e.g., entering the number to be called) under the supervision of the Scheduler which will keep the workload within bounds. It also informs the Workload Estimator of driver actions, such as being engaged in a phone conversation.

User interface and applications

The user interface makes it possible for the user to control the GIDS system and a few embedded applications, specifically telephone, tuner and amplifier. It also enables the GIDS system to address the user (Piersma et al., 1991).

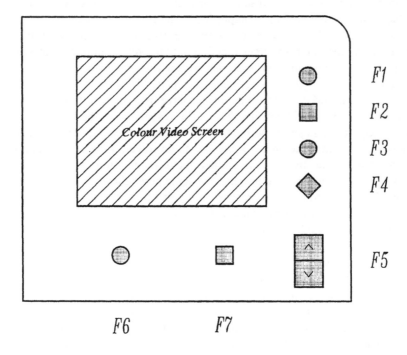

Figure 8.3 The layout of softkeys and a fixed function rocker switch around the GIDS screen, mounted within easy reach to the right of the steering wheel

The interface incorporates:
- high resolution, high illumination colour screen with four soft keys to the right of the screen and two at the bottom and a fixed-function rocker switch (see Figure 8.3);
- spoken messages;
- voice command.

A soft key is a key that has a variable function depending on the context, which, in this case, is an item on the adjacent screen. The soft keys to the right of the screen drive the user menus. The soft keys below the screen serve as an alternative to voice commands for picking up and hanging up the phone. The rocker switch tilts upwards for viewing previous screens and downwards for the repetition of spoken messages.

A set of prerecorded (digitized) messages is available for presentation. These messages are presented by a console speaker mounted below the screen.

A head-mounted microphone is used for verbal commands to the GIDS system and for phone conversations. Voice command uses an echo/cancel protocol: each command recognized is echoed to the user, allowing time for it to be cancelled by the user, before executing it.

Menus

While the car is standing still, a menu structure is available on the interface screen. The soft keys to the right of the screen are used for selection of items from this menu; a keyboard for more complex data entry is also available. The following menu structure is supported. Note, however, that the actual language used in the prototype is Dutch:

Hoofdmenu	Main Menu
Navigatie	Navigation
Aan/Uit	On/Off
Telefoon	Telephone
Neerleggen	Put on hook
Opnemen	Take off hook
Opbellen	Phone number
Nummer invoeren	Enter number
Radio	Radio
Aan/Uit	On/Off
Systeem	System
Ondersteuning	Support
Gaspedaal uitvoer Aan/Uit	Accelerator pedal output On/Off

Stuurwiel uitvoer Aan/Uit	Steering wheel output On/Off
Koers Aan/Uit	Lane keeping On/Off
Snelheid Aan/Uit	Speed On/Off
Persoonlijke gegevens	User profile
Invoeren persoonlijke gegevens	Enter profile
Onderricht	Tutorials
Halt systeem	Halt system
Halt systeem? Ja/Nee	Really halt system? Yes/No

Under 'Enter profile' the user's gender, age and experience can be entered. 'Tutorials' was designed to allow the PSALM module (not yet implemented) to be interacted with.

Telephone
The telephone in the GIDS prototype allows both incoming and outgoing calls. Calls from the car may be made using menus or voice commands. Incoming calls can also be accepted by using voice or manual response. Similar options are available for terminating calls. As mentioned above, incoming calls will not necessarily ring immediately, since workload adaptation may require postponement of the presentation of the ringing signal. Otherwise the telephone operates as a standard telephone, particularly with regard to the auditory signals presented.

Tuner
The tuner can be switched on and off. Station is predetermined.

Voice Command
The following commands are available for the voice control and again the actual language is Dutch.

Radio Aan	Radio On
Radio Uit	Radio Off
Geluid Harder ... Stop	Volume Up ... Stop
Geluid Zachter ... Stop	Volume Down ... Stop
Navigatie Aan	Navigation On
Navigatie Uit	Navigation Off
Waarom-Een-Waarschuwing?	Why-A-Warning?
Bel <*rij cijfers*> Op	Phone <*list of digits*> Up
Telefoon Opnemen	Pick Up Phone.
Telefoon Neerleggen	Hang Up Phone.

Following a command concerning the audio volume, the system slowly and continuously increases or decreases the volume (as requested) until the driver says "stop". In this case the echo/cancel protocol is ommitted from the "stop" command.

Whenever a message is presented to the driver via one of the active controls, a spoken message is available on request ("Why-A-Warning?") that explains the system's behaviour in terms of the objects sensed and the conclusions drawn. The rationale behind this function is as follows. If the reason for a warning is not immediately obvious to the driver, the further explanation provided by the why-a-warning function will either help the driver to recognize his or her error, or to accept the fact that the system's sensors were unreliable in that particular instance, although the system's reasoning was reliable. It is assumed that the why-a-warning function will help to maintain the driver's trust in the reliability of the system.

To dial a specific phone number, the driver enters the respective one-digit numbers one by one between "Phone" and "Up".

Adaptive volumes and audio switching
If the radio is playing, the effect of using the telephone is to switch the radio off until the phone call has been terminated. The master volume output is adjusted to the noise level in the car, so that GIDS messages and the radio can easily be heard in virtually all circumstances, but not too loudly when the level of ambient noise is low.

8.6 Control support: active car controls

The active car controls are the accelerator pedal and steering wheel (Chapter 7.3). Strictly speaking these are part of the user interface. The steering wheel is used for suggesting steering actions to the driver by discrete pulses. The accelerator pedal is used in two ways. One is by discrete changes in counterforce, the other by coupling the counterforce continuously to a variable of the driving task (Färber et al., 1990), in the GIDS prototype, the degree to which the maximum velocity allowed is exceeded. See Sections 8.8 and 8.9 for details concerning hardware.

8.7 The GIDS system hardware

This section describes the hardware used for the various software modules discussed so far, and the hardware adopted for the user interface. Table 8.1 shows the set of ISA-bus PCs used, the allocation of software modules to these processors, the ISA plug-in boards used, and an overview of the way these boards are connected. Table 8.2 shows the additional hardware used. Table 8.3 specifies the required software platforms. Figure 8.4 shows a diagram of the components, presented below where the ICACAD version is described (Section 8.9).

Table 8.1 GIDS system prototype processor hardware

Module (Processor) - ISA bus card	Connections
1 Speech (386/33-16M+HD)	
- MR8 Speech recognizer (Marconi Speech & Information System, Portsmouth, UK)	<- Microphone Amplifier
- PCRadio (Dutch PTT)	-> Audio Control Modules
- Soundblaster	-> Audio Control Modules
- Ethernet adapter	
2 Dialogue Generator (386/33-8M)	
- VDO screen driver	-> screen (glass fibre cable) <- CGA signal
- CGA graphics card (MetaGraphics compatible)	-> screen driver
- Sirex Voicecard (Sirex, Enschede, NL)	<- noise sensor microphone
- Modified Hayes Modem (custom built)	-> LLIN/ Carvox 4000 <- Microphone Amplifier
- Roland MPU-IPC-T MIDI	-> Audio Control Modules
- Ethernet adapter	
3 Scheduler and Workload Estimator (486/33-8M)	
- Ethernet adapter	
4 Manoeuvring and Control Support Module (486/33-8M)	
- Ethernet adapter	
5 Navigation Module (386/33-8M)	
- Ethernet adapter	
6 Personalized Support and Learning Module (not implemented) (386/33-8M)	
- Ethernet adapter	
7a Instrumentation Computer ICACAD (386/27-4M)	
- Ethernet adapter	
- ICACAD specific hardware outside GIDS system	
7b TRC-Simulator (IRIS-VGX)	
- Ethernet adapter	
- TRC-Simulator specific hardware outside GIDS system	

Table 8.2 Other hardware

Component	Connections
- Head-mounted microphone	-> Microphone amplifier
- Noise sensor microphone	-> Sirex voice card
- Audio Control Modules	-> Console amplifier
(Niche, AudioControlModule1)	-> Entertainment amplifier
(R. Jones Marketing, Northridge, CA)	<- Soundblaster
	<- PCRadio
	<- Hayes Modem
- Microphone amplifier (custom built)	<- Head-mounted microphone
	-> Hayes modem (+66 dB)
	-> MR8 (+0 dB)
- Console amplifier	<- Audio Control Modules
	-> Console speaker
- Console speaker	<- Console amplifier
- Entertainment amplifier	-> Entertainment speakers
- Entertainment speakers	<- Entertainment amplifier
- VDO enhanced LCD-CGA screen 9.5 x 12.8 cm	<- VDO screen driver
- Carvox4000 car telephone	-- Local Line Interface
- Local Line Interface	-- Modified Hayes Modem

Table 8.3 Software platform required

- DOS 3.3 or higher
- EMS driver (for 1, 2 and 4 only)
- Ethernet packet driver
- MetaWindows Kernel (for 2 only)

Descriptions

Figure 8.4 shows the mapping of the processor numbers as assigned in Table 8.1 onto the software processes (modules) as shown in Figure 8.1, for both versions of the prototype. Figure 8.5 presents all GIDS relevant hardware in the ICACAD version of the system in a single diagram. Next the various pieces of hardware will be briefly described.

— VDO-enhanced LCD-CGA screen 9.5x12.8 cm
For displaying pictograms in full daylight while driving and menus while standing still a special high contrast high illuminance screen by VDO is used.

— Special dashboard keypad
Along with the screen a special dashboard extension containing a set of softkeys aligned with the screen was devised. The allocation of keys is shown in Figure 8.3. The keys are interfaced through the Car Body Interface module.

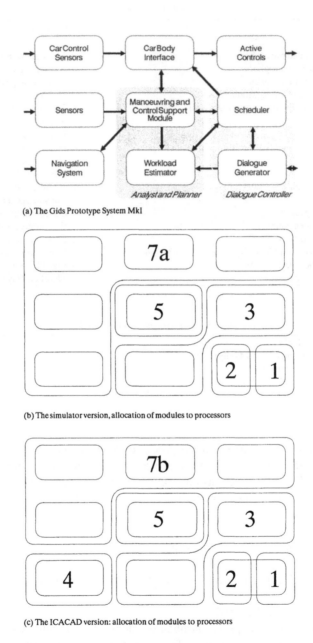

(a) The Gids Prototype System MkI

(b) The simulator version, allocation of modules to processors

(c) The ICACAD version: allocation of modules to processors

Figure 8.4 Allocation of the modules in the GIDS prototype system to the processors specified in Table 8.1

−   Speech input

A Marconi MR8 Speech recognizer is used. This allows recognition of keywords from continuous speech independent of individual speaker's characteristics.

−   Speech output

Two devices are used here, one to produce the speech output and another to control the sound level in order to take the background sound levels in the vehicle into account. This module allows the generation of verbal messages from sequences of segment labels which are linked with digitized speech segments. A stand-alone speech sample editor can be used to cut the digitized utterances into the required segments and link them to the appropriate labels. Apart from speech segments sound signals may also be used.

−   Soundblaster

The Soundblaster is an 8-bit sampler ISA-bus card that provides enough quality for producing speech and sound output using prerecorded sound samples. A female voice, sampled at 10 kHz, was used.

−   Sirex voicecard

The Sirex voicecard is yet another sampling ISA-bus card. It was chosen because of the availability of Fast Fourier Transform software to be used as a noise sensor. It provides the data necessary for adjusting the sound output levels of the GIDS system.

−   Car Telephone

A Carvox4000 Car Telephone with Local Line Interface (LLIN) and a Modified Hayes-compatible Modem have been installed. Using a Hayes modem with tapped audio signals allows computer control over incoming and outgoing calls. The LLIN allows connection of a regular Carvox4000 to a Hayes-compatible modem.

−   Radio

A PCRadio ISA-bus card is used. This provides for a prototypical radio under computer control.

−   Miscellaneous audio devices
    −   Microphone amplifier. A custom-designed amplifier is used to adjust the output level of the standard head-set microphone for the MR8 for input to the Modified Hayes Modem;
    −   Niche ACM1 Audio Control Module with Roland MPU-IPC-T MIDI interface. Most audio signals from the GIDS system pass through the audio control modules that allow each audio level to be adjusted;
    −   Speakers and amplifiers.

A GIDS vehicle must provide for three audio channels: Right-entertainment, Left-entertainment and System. In the simulator prototype entertainment is from behind the driver and system sound is from a speaker close to the GIDS screen, in the ICACAD entertainment is from the right and left ends of the top of the dashboard for pragmatic reasons.

The processors communicate using a thin Ethernet segment and a public domain packet driver plus customized communications interface.

8.8 Implementing GIDS in a vehicle 1: the TRC simulator

Installing the GIDS prototype system in a driving simulator like the one at the Traffic Research Centre (Chapter 9) is not a difficult task as far as the active controls are concerned, because accelerator pedal and steering wheel are already completely computer controlled. It simply involves customizing the software to modulate the forces on accelerator pedal and steering wheel so as to exhibit the proper feedback required by the GIDS system. Given the availability of the Small World Simulation software (see Chapter 9) the connection of GIDS to the simulator via Ethernet is relatively straightforward too. The actual job consisted of implementing the required sensory functions and adapting the simulator data representation to the representations used in the GIDS system. As the GIDS system initially uses the standard UDP/IP protocol to accommodate the connection to the simulator and, in future extensions, to other machines supporting UDP/IP (e.g., Unix machines), no major task was involved. The simulator interface to GIDS provides a Car Body Interface, Sensors, and a Navigation System.

A major effort, however, was required to customize the dashboard hardware. The GIDS prototype requires its screen with special function keys to be mounted to the right of the steering wheel, with a speaker below the screen. In the TRC simulator's BMW-518, most of the ventilator console hardware had to be removed to make way for the GIDS screen and surrounding frame holding the buttons. A speaker was mounted below this display. The entertainment speakers were mounted on the parcel shelf at the back of the car.

8.9 Implementing GIDS in a Vehicle 2: the TNO Instrumented Car for Computer Assisted Driving (ICACAD)

The Instrumented Car for Computer Assisted Driving (ICACAD) at the TNO Institute for Perception was built for road studies involving advanced support systems in general and GIDS in particular. It meets the following criteria:

- Handles like a regular passenger car ('feel');
- Sufficient space to carry instrumentation equipment including GIDS system computers and hardware;
- Experimenter (usually a certified driving instructor) is able to intervene in driver action;
- Allows implementation of a set of sensors (see below) for measuring driver behaviour and supporting the GIDS system;
- Allows implementation of active controls;
- Dashboard with room for extensions, to allow installation of GIDS screen and buttons;
- Data recording facilities;
- Allows interfacing of sensors and user interface elements to GIDS;
- Support appropriate power supplies;
- Room for technician to sit in the back while operating system hardware.

The following car and hardware were used:

- Dodge RAM Van, with extra windows in the back;
- Extra 24V/75A generator fitted in engine compartment;
- 500W/220V power supply for ICACAD and GIDS hardware;
- 1000W/220V power supply spare for extensions or specific experimental set-ups, e.g., video registration equipment;
- Two 24V/100Ah batteries for operation with engine switched off;
- Emergency button that shuts down engine, mounted on dashboard in front of passenger (experimenter) seat;
- Additional braking pedal mounted on floor in front of passenger seat;
- Two servo-motors for accelerator pedal and steering wheel;
- Technician's chair in the back, in front of instrumentation computer (see below).

The following sensors are available, although not all are currently used by GIDS (see Section 8.2):

- steering wheel position ($\pm 90°/540°$, $\pm 0.1\%$);
- steering wheel torque (0-50N, $\pm 2\%$);
- accelerator pedal position (pedal range, $\pm 0.5\%$);
- accelerator pedal torque (0-500N, $\pm 1\%$);
- braking pressure (0-80 At, $\pm 1\%$);
- lateral position (only when road has delineation) (-1 - +2.5, ± 2cm);
- lateral acceleration (± 1g, $\pm 1\%$);
- longitudinal speed (0-141 km/h, $\pm 2\%$);

- longitudinal position (30m to infrared beacon > 25cm (speed dep.) ≤ distance/sampler duration);
- longitudinal acceleration (± 1g, ± 1%);
- yaw rate (± 100°/s, ± 1%);
- roll rate (± 100°/s, ± 1%);
- usage of indicators;
- usage of car lights;
- headway sensor by way of a datalink to lead vehicle, that frequently sends position and speed. TNO's other instrumented car, ICARUS (Instrumented Car for Road User Studies) is equipped for this purpose with an infrared beacon sensor (see below) and remote RS232 interface;
- infrared (IR) beacon sensor. Both ICACAD and ICARUS have been equipped with IR sensors that trigger when IR reflecting surfaces are passed. The sensor is used to frequently calibrate the longitudinal position of the vehicle.

For data recording and interfacing a 386 AT computer is used in ICACAD with a multifunction I/O ISA plug-in card connected to custom hardware interfaces with sensors and active controls. It has four major functions:

- Record driver behaviour and GIDS system behaviour during experiments;
- Interface the GIDS system to the car (Car Body Interface and Sensors);
- Dead reckoning (this is interfaced to the Navigation system);
- User Interface for on-board technician, allowing communication to and monitoring of the GIDS computers, which do not have screens and keyboards of their own.

Installing the GIDS system

To accommodate the GIDS system hardware, a rack was mounted in the back of the car to hold the GIDS computers and other hardware, as schematically shown in Figure 8.5 and listed in Tables 8.1 and 8.2.

The GIDS screen console was added to the right of the original dashboard. The console speaker was mounted in an already available compartment below the screen. The entertainment speakers were mounted right- and leftmost on top of the dashboard. The ICACAD Instrumentation computer was connected to the GIDS Ethernet.

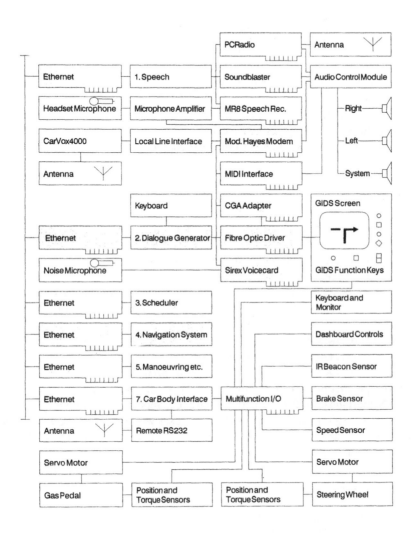

Figure 8.5 The GIDS relevant hardware in the ICACAD vehicle (see Section 8.7 for a description of the GIDS system hardware components and Section 8.9 for an explanation of the ICACAD specific hardware)

8.10 Conclusion

This completes the description of the two versions of the GIDS prototype. It underscores the feasibility of the general GIDS concept. In the process of realizing an instantiation of such a concept one must be pragmatic, and specify, design and implement a rather modest representation of the GIDS concept. However, the architecture proposed in this chapter is general enough to allow extensions in various directions in the future.

One aspect of the prototype has not yet been treated in detail: its domain of operation, the Small World, the environment for which the GIDS system behaviour is defined. This will be the subject of the next chapter.

References

Färber, B., Färber, Br.A., Godthelp, J., & Schumann, J. (1990). *State of the art and recommendations for characteristics of speed and steering support systems.* Deliverable Report DRIVE V1041/GIDS-CON1. Haren, The Netherlands: Traffic Research Centre, University of Groningen.

Groeger, J.A., Kuiken, M.J., Grande, G.E., Miltenburg, P.G.M., Brown, I.D., & Rothengatter, J.A. (1990). *Preliminary design specifications for appropriate feedback provision to drivers with differing levels of traffic experience.* Deliverable Report DRIVE V1041/GIDS-ADA1. Haren, The Netherlands: Traffic Research Centre, University of Groningen.

Piersma, E.H. (1990). *The first GIDS prototype: Module definitions and communication specifications.* Deliverable report DRIVE V1041/GIDS-DIA1A. Haren, The Netherlands: Traffic Research Centre, University of Groningen.

Piersma, E.H. (1991). Real time modelling of user workload. In Y. Quéinnec & F. Daniellou (Eds.), *Designing for everyone* (Vol. 2, pp. 1547-1549). London: Taylor and Francis.

Piersma, E.H., Verwey, W.B., Michon, J.A., & Webster, E. (1991). *Prototype specifications.* Deliverable Report DRIVE V1041/GIDS-GEN03. Haren, The Netherlands: Traffic Research Centre, University of Groningen.

Smiley, A., & Michon, J.A. (1989). *Conceptual framework for generic intelligent driver support.* Deliverable Report DRIVE V1041/GIDS-GEN01. Haren, The Netherlands: Traffic Research Centre, University of Groningen.

Verwey, W.B. (1990). *Adaptable driver-car interaction in the GIDS system: Guidelines and a preliminary design from a human factors point of view.* Report IZF 1990 B-12. Soesterberg, The Netherlands: TNO Institute for Perception.

Verwey, W.B. (1991). *Towards guidelines for in-car information management: Driver workload in specific driving situations.* Report IZF 1991 C-13. Soester-berg, The Netherlands: TNO Institute for Perception.

Chapter 9
GIDS Small World simulation

Wim van Winsum, Peter C. van Wolffelaar

9.0 Chapter outline

This chapter describes the Small World Simulation facility which was built in the course of the GIDS project with a dual purpose in mind: (a) to have a research tool to allow rapid prototyping and testing of various GIDS functions and components and (b) to have a simulation facility with a complete GIDS system installed for behavioural studies that would be too laborious or too dangerous to perform in the real world. First, in Section 9.1 we describe the functionality of the Small World Simulation in general terms. This is followed in Section 9.2 by a detailed description of the major elements of the simulator, the graphical simulator, the vehicle dynamics model, the graphical environment, and the traverser. The traffic environment and the way it is interfaced with the GIDS simulation is the topic of Section 9.3. Running the simulator in a particular configuration requires a set of special utilities, including a network editor, a car initializer, and a scenario editor. These are described in Section 9.4. Next, the suitability of the Small World Simulation as a test environment for GIDS will be considered in Section 9.5. Finally some experimental issues and the simulator's potential for data collection are discussed in 9.6.

9.1 Functionality of the Small World Simulation

The Small World Simulation consists of driving in a dynamic traffic environment in the simulator at the Traffic Research Centre of the University of Groningen. This simulator allows the driver of the simulator car to move freely through a network of roads as if driving in the real world. On the way, several other vehicles are met that react to the simulator car 'naturally' and in real time. These other vehicles all

175

have their own intelligence which allows them to make their own decisions, dependent on the behaviour of the simulator car, the other computer-controlled vehicles, the presence of traffic signs, and traffic lights, and the layout of the network of roads. In addition to a dynamic traffic environment the simulated world consists of buildings, traffic signs, road markings, and traffic lights which can be controlled independently through a number of different strategies. The noise the driver is hearing conforms with the pressure exerted on the accelerator pedal, the velocity of the car and a side wind.

The system is designed such that driving through the simulated world is as realistic as possible. The simulator car behaves like a real car, the other traffic moves and behaves as if it is real traffic, the perspective and lighting conditions as well as the casting of shadows are quite realistic, and the noise heard while driving in the simulator car strongly resembles the sound while driving in real traffic. However, since the simulator is fixed-base, the driver lacks the locomotor sensations experienced in real driving.

Figure 9.1 Computer-generated image of road environment with interacting cars

The user is free to specify any kind of road net as long as the physical characteristics of the Small World conform with the specifications described in Groeger et al. (1990) and in Chapter 5. For the simulator experiments described in Chapter ·10, a fairly complex and large road net was designed, which may be regarded as an extended small world. The Small World consists of a subset of real-world situations in order to reduce the complexity of the normal driving task. It contains straight road sections, curves with different radii, T-junctions, crossroads, and roundabouts. The roads are dual-lane and allow two-way traffic. One-way roads are indicated by T- signs. Right of way either complies with standard traffic regulations or is controlled by signs. Interactions between traffic participants at intersections can also be controlled by traffic lights. All traffic participants are vehicles which behave according to standard rules. These rules are described in Van Winsum (1991).

The simulator offers a number of advantages over testing in the real world:
- Since the positions, velocities, etc., of all traffic participants are represented in the simulator, important sensor information is available that would be much harder to obtain with current technology in a real world test situation. This information is vital as input for the Analyst/Planner, making the simulator functionally equivalent to a Sensory Integrator and Car Body Interface for the GIDS system (see Chapter 8).
- All kinds of constellations of situations, or event sequences, can be tested. This allows analysis of message structures and dialogues that cannot easily be tested in a real-world experiment.
- The response of GIDS to manoeuvres that are too dangerous to test in real-world driving can easily be tested in the simulator.
- Situations can be brought under experimental control. This is important for the comparability of subjects' results, since all subjects can encounter exactly the same situations.
- Data collection in experimental situations is greatly facilitated by the simulator since all relevant performance measures are continuously available.

9.2 The structure of the simulator

Physically, the driving simulator consists of a car body and its driving controls, a large graphical projection screen showing the road environment, and some digital computers running the software for graphical projection and the simulation models. Functionally, the system consists of two main components: the 'classical' driving simulator with its car cabin, sound generation and graphical projection system, and an additional simulated interactive traffic environment with its logical road structure and its 'intelligent' vehicles. The latter enables the driver to interact realistically with other road users while driving in the simulated road network.

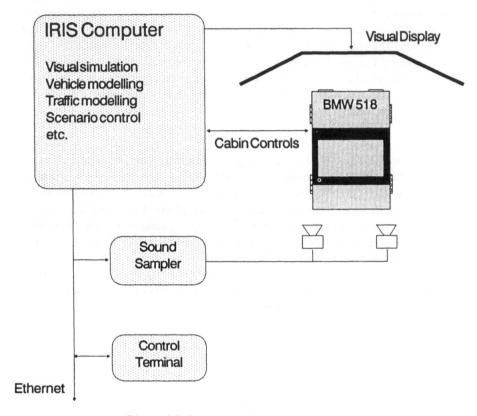

Figure 9.2 Overview of system components

Graphical simulator

The heart of the simulator is a Silicon Graphics 'IRIS VGX' graphical workstation located at the Traffic Research Centre, University of Groningen. This computer system allows real-time calculations and graphical display of dynamic three-dimensional scenes. The IRIS is directly linked to the simulator car body for the exchange of car control data from the steering wheel and pedals, and is connected via an Ethernet port to other peripheral hardware and to the GIDS prototype system. In order to obtain a smooth visual projection of the outside view, the system generates the updated outside projections at the highest possible rate, limited mainly by the capacity of the graphical hardware. In a continuously running program loop, several sequential processes are repeated for each picture update. First, positions of car body controls are determined by sensing inputs, and the responses of the simulated car are computed by car model software running on the IRIS. Next, the resulting speed and heading of the car are processed by a traverser module which determines the new position in the simulated world. At this stage the interactive traffic simulator module, to be described below, integrates the movements of the simulator car driver into its own traffic processes and determines new world positions for all traffic participants. Finally, the updated outside world is projected on the screen as viewed from the position of the driver in the simulator cabin and the cycle will start all over again. The system thus enables a real-time driving environment to be displayed with a screen refresh rate of 20 Hz and a system response delay of 50-60 milliseconds.

Car body

The simulator car cabin consists of a modified BMW 518 cabin containing a steering wheel, clutch, accelerator pedal, gear lever, brake pedal, console switches, and turning indicators. An additional colour LCD screen which is part of the GIDS system has been built into the dashboard, just to the right of the normal display panel. A servo-motor system is attached to the steering mechanism, in place of the original assembly, to produce dynamic forces on the steering wheel. Besides simulated driving forces, as calculated by the car model software on the IRIS, the servo-motor also delivers the GIDS warning torque signals to the steering wheel. A similar servo-system is attached to the accelerator, enabling GIDS to generate signalling forces on the accelerator. All positions of cabin levers and switches (gear box, light, horn, contact, etc.) are sensed by the IRIS at screen refresh rate, and are thus continuously available for use by the car model and for data collection. To increase the functional realism of the simulator, driving sounds are generated by a digital sound sampler controlled by the car model on the IRIS and presented by a stereophonic loudspeaker system in the front of the cabin. While driving two sources of sounds are generated: the engine noise with its changing pitch and volume depending on engine revolutions; and engine power and wind noise, depending on driving speed.

Vehicle dynamics model

The driving dynamics of the simulated vehicle are controlled by the car model module in the simulation software. At the start of each frame-refresh cycle an update is calculated for the car's instantaneous driving speed and heading. Internally the dynamic behaviour is simulated by a model of the car dynamics, based on transfer of energy between engine and wheel axis, and mediated by clutch, gearbox, and differential. Input signals from the car body are the actual positions of the accelerator, clutch, brake, gear stick, and steering wheel. Outputs are engine rotations per minute (rpm), driving speed and wheel heading, all of which are subsequently used in the interactive road traffic model for calculations in the traverser functions. The output of engine rpm and driving speed are also used, together with accelerator position, to calculate engine and wind sound parameters, which are directly sent to a digital sound sampler.

Graphical environment

The graphical display through the windscreen shows the road, traffic signs, vehicles, and buildings in the perspective seen from the driver's position. The network parameters are contained in a data file that describes the dimensions and the layout of the road structure (see Section 9.4, 'network editor'). Since the capacity of any graphical system is limited, the degree of complexity of graphical objects essentially determines the possible frame-refresh rate. In practice a balance must be found between the desired level of detail in the scenery and the required minimum screen update rate. As a refresh rate of less then 15-20 frames per second is unacceptable in a real-time driving environment, it is inevitable that the degree of complexity in the simulated scenery will be quite restricted. Therefore in this simulator detailed representations of graphical objects have generally been limited to parts that are essential for traffic interactions. Thus the appearance of buildings is kept simple since their only function, in terms of traffic behaviour, is to create visual occlusion. Cars are constructed from a few polygons but they do display brake and signal lights. On the other hand, their movement patterns, as determined by the interactive traffic module, are very realistic.

Signs are shown in detail as far as is required for detection and recognition. In contrast the road structure with its surface markings has been computed to more accurate detail and smooth curvatures so as to obtain maximum realism for visual guidance. The IRIS graphical hardware also enables visual contrast to be simulated giving all object surfaces the appearance of being illuminated by a distant source – namely the sun – from a particular position in space.

At the present stage of development the software allows the generation of two-lane road systems, various traffic signs and (groups of) buildings situated along the road segments.

Traverser

The instantaneous speed and heading of the vehicle are fed into the Traverser. This module calculates the lateral displacement (with respect to the right side of the road) and the longitudinal displacement, the heading of the car, and its position in terms of the coordinate system of the road network. It connects the simulator car to the road network, checks which path is selected if the car is negotiating an intersection and performs a number of other checks to maintain accurately the position of the simulator car with respect to other traffic. It uses a number of dead-reckoning techniques in this process. The output of the Traverser are the coordinates of the centre point of the car and the heading of the car. These values are transferred to the graphics system. Thus the Traverser is the intermediary between the graphical simulator and the traffic environment.

9.3 The traffic environment

The traffic environment consists of a network of roads and a number of computer-controlled vehicles, each with its own sensors and intelligence moving through the network.

The static environment or network

The static environment consists of a network of roads with related traffic signs and traffic lights. This environment is derived from a parametric description of a network read from a file created by means of the network editor.

The network consists of a logical structure of three basic tables: a table with intersections, a table with paths and a table with segments. Figure 9.3 shows the relations between intersections, paths, and segments for a path (P1) and its counter-path (P2). Intersections are represented by S1, S2 and S3. A logical intersection is a point in the network coordinate system connecting two or more paths. An intersection is of a certain type, for instance, type *roundabout*. It can be controlled by traffic lights with a certain control strategy, and it contains a list of references to outgoing paths. This list is ordered such that the path connections to the intersections are counter-clockwise, that is, the next path in this list is always the first path on the right. A path is a logical connection between two intersections that always has one direction. Each path has a counter-path in the opposite direction. In the table, paths are grouped in pairs, such that a path has an even index number and its counter-path has an odd index number or vice versa. Each path contains a list of references to segments. This list is ordered such that the segments are in successive order. A path also contains information about right-of-way at the intersection at the end of the path, about whether entry into this path is allowed, and it contains a reference to a traffic light at the end of the path if there is one. Finally, a segment is a vector that is a representation of a part of a road. It can be either straight or curved and it is essentially undirected. The direction depends only on the path the segment is in. If the segment is straight, the two end points are given in coordinates. If it is curved, the segment also contains the necessary information on the curvature, such as the radius, the centre point of the arc, etc. A segment is of a certain type which determines the width of the road and the number of lanes.

The three basic tables are processed to derive other tables which contain more information on how the computer-controlled vehicles should traverse the network. An ordered list with references to vehicles is attached to every path. The list ordering reflects the ordering of the vehicles on the path. In this way the simulator car and the computer-controlled vehicles are connected to the road network.

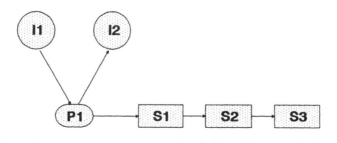

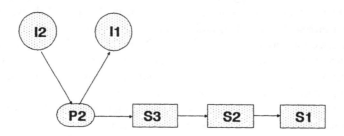

Figure 9.3 Structure of the road network

Sensors

The computer-controlled vehicles have several sensors at their disposal. The sensors constitute the interface between the intelligence and the network. They consist of perceptual functions to search for objects and to inspect the layout of the roads. One class of sensors enables the computer-controlled vehicles to check the presence of vehicles behind, in front, approaching from the opposite direction, and approaching from left, right, or straight ahead at the next intersection. By means of these sensors the vehicles have access, for instance, to their distance from other vehicles and their velocity. Another class of sensors enables the computer-controlled vehicles to determine the distance from an intersection, to detect if there are traffic lights ahead and what their status is, to inspect the layout of the road and to detect the presence of traffic signs. For experimental purposes, the range of sensors can be adjusted, or they can be switched off altogether.

Intelligence

The vehicles control their speed and steering behaviour by applying a number of hierarchically structured decision rules. A decision rule evaluates the sensor infor-mation, checks the presence of other objects in the vicinity of each vehicle, and uses its distance from these objects, their velocity, and a number of other properties of the objects to calculate a required velocity or a required lateral position. In this calculation a descriptive model of human driving is used. This model has been presented in detail in Van Winsum (1991).

There are decision rules for car following, keeping speed limits, velocity control at the entry to curves, in curves and intersections, depending on the next path and the layout of the network, lane changing, negotiating intersections and lane chang-ing on roundabouts, collision avoidance, emergency braking, overtaking, being overtaken, responding to traffic lights, checking signs for right of way, and others. Any higher level rule may invoke several lower level rules. For example, if the ve-hicle is close to an intersection, the high-level rule "negotiate intersection" is ap-plied. This rule, in turn, applies several other rules. The rule checks if there is a road on the left. If there is such a road and there is traffic approaching from the left, right of way rules for traffic from left are applied. Also, right of way signs are dealt with as well as traffic lights. Typically the various rules evaluate the situation the vehicle is in and then produce a maximum and minimum acceptable velocity, or a required change in lateral position. Since a number of rules may be applied simul-taneously, the output of all invoked rules is evaluated by a separate function that prescribes a velocity and lateral position. The prescribed velocity and lateral posi-tion constitute the input for a process which produces a new velocity and steering wheel angle.

All these rules serve three subgoals of the computer-controlled vehicles, namely, to avoid collisions, not to endanger other vehicles, and not to disrupt the traffic flow. At a higher level there is the navigation goal. At this level the vehicles either select a random route or their route is predetermined according to the rules of the road, but even if they follow a random route, general rules, such as those governing turning into one-way streets, will not be violated.

The velocity and steering wheel angle output by the intelligence are fed into a process, the traverser (see Section 9.2), which determines the coordinate position and heading of the vehicle. These values are then made available to the graphical system. The vehicle does not necessarily follow an ideal path, but is free to deviate from the 'ideal' course in the centre of the right lane. However, the longitudinal distance, measured from the start of the path, is normalized to the path along the centre of the right lane. The integration of the 'classical' driving simulator and the traffic environment constitutes the functional architecture of the simulator. This is represented in Figure 9.4.

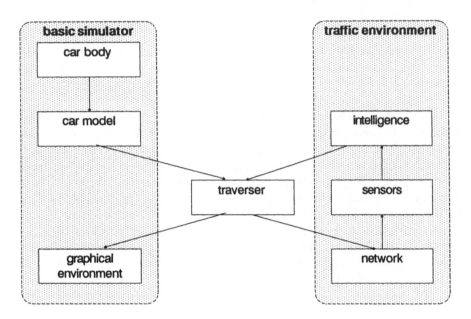

Figure 9.4 Functional architecture of the simulator

9.4 Utilities

In order to operate the simulator requires initial input. When the simulator program is started, a geometric description of the road network is read from a file. This file also contains a description of positions of traffic lights, signs, and buildings. The initial positions of all vehicles, including the simulator car, the routes (if any), properties of the vehicles and of their virtual 'drivers' are read from another file. If scenarios are to be used, descriptions of the scenarios are read from a third file. Scenarios discribe events that happen during the course of a simulator run. These files can be created by a package of utility programs which runs on a PC. The utilities are vital for experimental flexibility and they enable the experimenter to design a network which suits the requirements of any specific experiment, to assign properties to the vehicles, and to control the type of encounters that may occur between the simulator car and other traffic during the course of the experiment. These programs are integrated into one package, with graphics screens and mouse control.

Network editor

The network editor enables the user to design a network of roads. It has facilities for:

— adding and deleting intersections and roundabouts;
— adding and deleting paths between intersections;
— adding and deleting straight and curved segments;
— positioning traffic lights at intersections and implementing a number of control strategies;
— positioning a number of different types of traffic signs;
— positioning buildings of several types;
— scaling and zooming in on (parts of) a network;
— reading networks from file and writing networks to file.

Car initializer

The car initializer is a program which allows the user to:

— position vehicles in a network (including the simulator car);
— assign routes to vehicles (including the simulator car);
— assign properties to vehicles, such as wheel base, reaction time of driver, etc. (including the simulator car);
— activate or deactivate sensors for individual vehicles or all vehicles simultaneously;
— activate or deactivate decision rules for individual vehicles or all vehicles simultaneously;
— read vehicle properties from file and writing to file.

Scenario editor

The scenario editor allows the user to specify events that are initiated when the simulator car arrives at a certain position. The user is able to specify for every scenario:

— position of simulator car at start of scenario;
— position of simulator car at end of scenario;
— computer-controlled vehicles involved in the scenario;

–   reassignment of properties to the involved vehicles, e.g., sensor activation, rule usage, and route to follow during the scenario;
–   reading of scenarios from file and writing to file.

9.5 The Small World simulation as a test environment

The link between the simulator and GIDS

All knowledge about positions of objects, distances between objects and the GIDS car, velocities of objects, and so on, are stored in the simulator. The computer-controlled vehicles have sensors to obtain information from a variety of objects. These vehicles use this information when applying their rules for speed control and lateral position control. In fact, the rules that are used by these vehicles are equivalent to the rules used by the Analyst/Planner of the GIDS system to determine whether the driver of the GIDS car is controlling both speed and lateral position within the acceptable range. If the simulator is linked to GIDS, the GIDS car gains access to these same sensors, but in this case the intelligence is, of course, the GIDS intelligence and not the intelligence of the computer-controlled vehicles.

These sensors transfer their information to the GIDS computer, where the Analyst/Planner then applies its intelligence to the incoming information. Since the incoming (simulated) sensor information is of high quality and reliability, testing the GIDS system can focus completely on the GIDS design and intelligence. As such the simulator acts as a reliable Sensory Integrator which provides GIDS with input of a quality that could not possibly be provided in real-world test situations. This is of paramount importance because a GIDS system can operate only if there is an input to process. The sensor information transferred to the GIDS computer consists of the velocity, distance along the path, heading angle, the lane, the lateral distance, the lateral velocity and the indicator status of vehicles in the vicinity of the GIDS car. These other vehicles may be:

–   lead vehicles;
–   rear vehicles;
–   oncoming vehicles;
–   vehicles approaching the next intersection from the left;
–   vehicles approaching the next intersection from the right;
–   vehicles approaching the next intersection from straight ahead.

Another type of sensor information refers to the static environment. It identifies the status and distance of a traffic light the GIDS car is approaching, and type and distance of a traffic sign.

The car model of the simulator transfers data on current velocity and wheel angle to a Car Body Interface. Information on car controls and active controls are also transferred to the Car Body Interface. The Car Body Interface, finally, transfers all this information to the GIDS computer, where it is processed by the Analyst/Planner.

Through its Dialogue Generator, GIDS sends information on accelerator counterforce and steering wheel vibrations to the simulator car cabin. The Dialogue Generator also sends information to the speech generators and to other applications in the simulator car cabin, thus completing the circle. These relations are summarized in Figure 9.5.

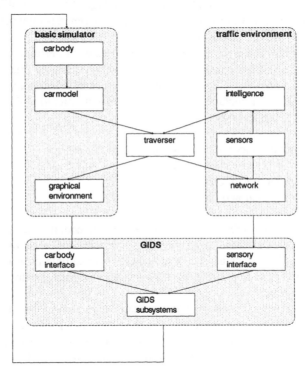

Figure 9.5 Relation between simulator and GIDS

Testing GIDS in a dynamic environment

It would be an unwieldy task to test all possible combinations of circumstances that call for a GIDS message or intervention. Therefore it is necessary to list the most interesting, critical interactions between the driver and GIDS for empirical testing. Furthermore, in order to test GIDS to the limit and to test the performance of the Scheduler, combinations of situations must be presented to the driver at the same time. This may be difficult to achieve in real traffic because of its intrinsically un-predictable nature and uncontrollability.

If the simulator is used as a test environment for the GIDS system, the simulator car acts as the GIDS car. In this case GIDS monitors the behaviour of the human driver. It also monitors the behaviour of the other vehicles through the sensory integrator (see Chapter 8). GIDS has, of course, no knowledge of the fact that these other vehicles are computer-controlled. Conversely, in the simulator, the computer-controlled vehicles can see no difference between the simulator car and the other computer-controlled vehicles.

Basically, there are two different kinds of circumstances in a dynamic traffic environment that call for a GIDS intervention or message. The driver of the GIDS car may make an error or drive unsafely, or else the other vehicles may behave in ways that endanger the (driver of) the GIDS car.

What is regarded as dangerous behaviour is a matter of definition. With respect to the intelligence of the computer-controlled vehicles and that of the Analyst/Planner, behaviour is regarded as dangerous when the current velocity and/or lateral position fall outside the acceptable range for the current situation. The current situation consists of a certain constellation of:

– positions and distances of other objects relative to the car;
– velocities of other objects relative to the velocity of the car;
– additional properties of other objects;
– properties of the car and the driver.

Additional properties of other objects include, for instance, the definition of a traffic sign and whether a traffic light is red or not. Another example of an additional property is the indicator status of a rear vehicle. An example of a property of the car is its braking capability and an example of a property of the driver is the individual's reaction time.

Thus, the presence and properties of some objects in the vicinity of the car determine the required behaviour of the car. Suppose there is a lead vehicle on the path the car is following, travelling a certain distance in front of the car and at a certain velocity relative to the car. Depending on the distance, the velocity of the lead vehicle, the velocity of the car, and some other properties of the car and the driver, there is a maximum acceptable velocity for the car. If the current velocity is higher than this acceptable maximum, the driver is considered to exhibit dangerous behaviour, because in the immediate future headway would be too short to avoid a collision if the lead vehicle suddenly decelerated.

(a) The GIDS driver behaves dangerously

Normally the computer-controlled vehicles behave according to rules that ensure safe behaviour. These vehicles also use rules for cases where other traffic does not behave according to safe rules. Since the simulator car is controlled by a human, this car may exhibit unexpected and unsafe behaviour. In that case the computer-controlled vehicles will act appropriately. Since these vehicles are "motivated to stay alive" they will try, by behaving appropriately, to neutralize the situation and, in the process also protect the driver of the GIDS car from becoming involved in an accident.

Also the GIDS system has to warn the driver in time or intervene by exerting a counter-force on the accelerator pedal. This implies that a number of GIDS messages and interventions may be tested by letting the driver explicitly make errors under controlled circumstances. The test facility also allows instruction dialogues of the personalized support and learning module, PSALM, to be tested. In fact, all the personalized support and learning module dialogues are tested by explicitly letting the driver make performance errors.

It would evidently be totally inappropriate to encourage a driver to exhibit dangerous behaviour during tests in real traffic. In the simulator, however, no harm is done by such tests.

(b) The other traffic behaves dangerously
An important task of the GIDS system is to ensure the driver's survival in a complex dynamic environment where all kinds of dangers may, often literally, lie around the corner. A number of tests can only be performed on the GIDS system by making the other vehicles in the vicinity of the GIDS car display unexpected and dangerous behaviour. Since sensors and rules can be switched off for the computer-controlled vehicles in the simulator, the production of dangerous behaviour can obviously be achieved more easily in the simulator than in the real world. Again, under real world conditions it would be too hazardous to ask drivers of other vehicles to display dangerous behaviour when they are close to the GIDS car simply in order to test the GIDS prototype, even if the real world test area would be closed for other free flowing traffic.

Furthermore, in the simulator test runs can be made of situations where the driver of the GIDS car makes no errors and where the surrounding traffic presents no danger for the GIDS car. Under these conditions GIDS should remain dormant. This can, of course, be tested in real world traffic, but the advantage of the simulator in this case lies again in the ability to control other traffic. A test facility of this kind enables testing of false alarm rates, which might be valuable for experiments on user acceptance and for fine-tuning the GIDS system. In this manner, criteria for message frequency can be tested and adjusted.

Complexity

In order to test the Scheduler, a number of situations that call for GIDS to act must be present at the same time. The number of situations that call for a message must be manipulated. In the simulator several potentially dangerous situations can be invoked at the same time. The Scheduler can be tested under situations of differing complexity. As the number of interactions with other traffic increases the number of potential GIDS measures increases. This allows complex timing issues to be studied in depth, possibly leading to optimization and better tuning of the Scheduler.

9.6 Experimental issues and data collection

Besides testing the functional correctness of the GIDS design and establishing whether GIDS works in a technical sense, an important question for evaluation studies of GIDS concerns the reaction of car drivers to GIDS, in addition to the reaction of GIDS to the drivers. These are questions concerning user acceptance, user interfaces, whether there is a shift to more risky driving, impact on workload, learning curves, impact on different driving experience groups, age groups, etc.

Comparability between subjects' results will be enhanced if all subjects encounter the same situations during the experiments. This can be achieved in the simulator because the situations the driver encounters can be predetermined by means of the scenario editor. A scenario is a constellation of situations that occur when the simulator car arrives at a certain point in the road network. If the car is, for instance, close to a certain intersection, a situation can be created such that the traffic light turns amber, the lead vehicle does not stop, a car approaches from the right with a given speed, and so on. This implies that traffic density and interactions with other traffic can be placed under experimental control. A scenario starts when the simulator car reaches a certain point. It ends when the car reaches another specified point within the network. In between, all interactions with the other cars are under full control. The scenarios used in the simulator evaluation studies of the GIDS system are described in Chapter 10.

This also allows experiments to be conducted on specific driver tasks, such as car following, negotiating intersections, overtaking and curve following, and on specific aspects of decision making, such as gap acceptance at intersections, speed choice on curved roads, speed choice under restricted visibility, speed choice while negotiating intersections with restricted visibility into sidestreets, and so on. This provides the possibility of evaluating GIDS interactions in special circumstances.

The route of the GIDS car and of computer-controlled vehicles can be determined in advance. Together with the database of the road network, this enables the simulator to send important navigation information to the GIDS system. The static environment can be designed so as to address a number of research questions. Test circuits can be easily and quickly made with straight roads, curved roads with any curvature and different kinds of intersections with all kinds of angles between connecting roads. Sight can be restricted at intersections or curved roads by introducing buildings that produce different amounts of occlusion.

This gives the experimenter considerably more control than is obtainable in real traffic. It also makes experimentation faster and more efficient because fewer subjects are needed to achieve a particular level of statistical power. In addition it greatly reduces the time required for preparation of experiments.

A large amount of performance data can be collected during a scenario or a complete test run, with any sampling frequency. Examples are time to collision, time to intersection, velocities, decelerations and accelerations, distances from objects, navigation errors, event codes, etc.

The possibility of replaying a test run constitutes another important advantage of the simulator. This enables the researcher to study interactions in any detail required. It also enables the researcher to measure other variables during the re-run than previously measured. In real-world experiments it is often hard to obtain reliable speed data because of sensitivities of the measurement apparatus to vibrations. Also cross-linking measured velocities with distances from other objects, which is necessary in order to compute time-based measures, is often very difficult, if not impossible, to establish in the real world. Since accurate measurement is entirely feasible in the simulator, this offers many advantages compared with real-world experiments.

The data for the simulator experiment as described in Chapter 10 are studied separately for the simulator car and for all other cars. For the other cars, only absolute time from the start of the test run, velocity, and distance to the simulator car are stored. The data for the simulator car include, besides the absolute time from the start of the test run, the current segment, the intersection the simulator car is approaching, whether the simulator car is on an intersection, the status of the traffic light ahead if there is one, the number of the current scenario, the velocity, the lateral distance, and the distance to the first intersection ahead.

References

Groeger, J.A., Kuiken, M.J., Grande, G.E., Miltenburg, P.G.M., Brown, I.D., & Rothengatter, J.A. (1990). *Preliminary design specifications for appropriate feedback provision to drivers with differing levels of traffic experience.* Deliverable Report DRIVE V1041 GIDS/ADA1. Haren, The Netherlands: Traffic Research Centre, University of Groningen.

Van Winsum, W. (1991). *Cognitive and normative models of car driving* Deliverable Report V1041 GIDS/DIA3. Haren, The Netherlands: Traffic Research Centre, University of Groningen.

Part III
Performance and perspective

Chapter 10
Evaluation studies

Wiel H. Janssen, Marja J. Kuiken

10.0 Chapter Outline

In this chapter we report the substance of the operational system and performance evaluation studies to which GIDS has been subjected thus far. These studies are by no means to be considered as exhaustive, but they rather form an initial empirical check of the ideas behind GIDS that could possibly reveal that these are blatant nonsense when really brought into the open.

Section 10.1 contains the reasoning behind these experiments. Section 10.2 deals with a first empirical study, which was carried out in the Small World Simulation facility described in Chapter 9. Section 10.3 concerns a second empirical study, which gives an overall comparison of the performance of the system in real-world traffic conditions in which the driver is either given or not given the benefit of GIDS support. This experiment was carried out in the city of Utrecht in The Netherlands. The concluding Section 10.4 provides a discussion and a summary of the results obtained in these evaluation trials to answer the question: How well does GIDS 'Mark I' stand up to scrutiny?

10.1 Introduction to the behavioural experiments

In the third year of the GIDS project evaluation studies were carried out to evaluate the impact of the prototype system with a view to safety, effectiveness, and impact on driver behaviour and workload.

The principal aim of these studies was to acquire knowledge about the performance of various categories of drivers who are operating under a variety of circumstances, in three basic conditions:

- driving without any support;
- driving with non-integrated applications;
- driving with (integrated) GIDS support.

Experimenting with GIDS

A system, such as GIDS, which contains so many new elements in so many new combinations, must be put to the test in as many ways as feasible before it can truthfully be offered as something of value to the driving population. A reasonable conviction must emerge that GIDS performs adequately in the situations for which it was designed and that a positive contribution to efficiency and safety can be expected.

Straightforward assessment of the performance of an in-vehicle system implies taking it out in real traffic and evaluating how driving with its support compares to driving without its support. This is one part of the evaluation of the GIDS prototype. However, driving in real traffic, even for an extended period of time, will yield no more than a sample of configurations in which GIDS should eventually play a role. Thus, while the situations encountered on the road may be real enough, many interesting configurations, particularly those of a somewhat critical nature, will occur infrequently or not at all. For this reason, experiments in a simulator study in which conditions were more strictly under control were carried out in parallel to the field experimentation.

This section will describe what we actually hoped to achieve with the experiment, what our expectations and doubts were, and how the experiments were planned such that they can indeed be said to constitute fair tests.

At the heart of the GIDS concept is that GIDS must be an integrated system, that is, it must be more than a collection of separate components that do not know of each other's existence. Therefore, the comparison central to the experiments should be between the integrated GIDS system – including the dialogue controller – and the collection of components for navigation, collision avoidance and vehicle control, all working in isolation. In addition, experimentation should also include a basic comparison with the situation of having to drive without any help at all, as it is not unlikely that there exist situations in which having no support at all is better than receiving several messages simultaneously from separate functions.

Since accidents are rare events, it is inconceivable that one would find, in tests spanning only a limited period of time, significant beneficial effects of GIDS support on the number of accidents or on their severity. Substitute measures, derived from driving performance itself, have to be used. Fortunately, many of these have already been derived; they have been described in the literature and have proved their use in experimentation. The specifics will be given below, in the more detailed description of the experiments (see also Janssen, Kuiken, Miltenburg, & Verwey, 1992; Kuiken & Miltenburg, 1993).

If we consider efficiency rather than safety, the performance of the navigation component of GIDS may be the most relevant feature to look at. This involves straightforward measures of success in reaching specified destinations, such as the time required to reach a destination and the number of errors made in getting there.

It is unlikely that a driver support system will have the same effects on everyone's behaviour. The fine-tuning to individual driving behaviour that GIDS in its present form already contains will not eliminate the basic fact that the better a driver, the less support will be needed. Thus, novice drivers should be expected to profit most from GIDS. In order to evaluate whether this is indeed the case and, if so, to estimate how large the profit might be, we have used both experienced and inexperienced drivers as our subjects. Another, similar, expectation is that older drivers would on average have more help from a support system than younger ones. For these reason we included both older and younger drivers as subjects.

If the variable 'age' is combined with the variable 'driving experience', four groups of drivers emerge. However, since older but inexperienced drivers have become a rare species we dropped this particular combination from the experimental design. Thus, we have searched for GIDS effects on the behaviour of young inexperienced, young experienced, and older experienced drivers.

Hypotheses

The essential comparison is therefore the one between the different GIDS conditions (no support / non-integrated support / integrated GIDS support) and their interactions with age and experience. The basic hypothesis, as far as actual driving performance is concerned, is that the overall improvement in performing the driving task will be largest with integrated GIDS support. Performance in the non-integrated condition should reveal small or possibly no improvements at all, relative to the condition in which there is no support. With respect to workload experienced by drivers the basic hypothesis is similar. Not only do we predict that workload will be higher under non-integrated than under integrated support, but we actually expect workload to be higher under non-integrated support than under no support at all.

How these conditions were realized requires no further explanation beyond saying that 'Non-integrated support' was created by simply disconnecting the dialogue controller from the separate components (see Chapter 8).

10.2 Small World simulation study

General

The simulator study allows for a much better controlled testing of the GIDS prototype than in the real world. An important advantage is also that one can create critical or dangerous situations in the simulator, which allows an in-depth study of both the performance of the GIDS system and the performance of the drivers. In Chapter 9 a detailed description is given of the simulated world, the simulator including car body and graphical environment, and the traffic environment. The aspects that are of particular interest for the present experiment are briefly mentioned below. A more detailed description of the simulator study is given by Kuiken and Miltenburg (1993).

The Small World

(a) The experimental route in the Small World
The experimental Small World route for the simulator study was based on the experimental route in the field experiment (see Section 10.3). The experimental route was approximately 8.5 kilometres long and, at an average speed of 30 km h^{-1}, took about 17 minutes to complete.

(b) Traffic in the Small World
Ten computer-controlled vehicles were moving about in the simulated road network, each with its own sensors and intelligence. These computer-controlled vehicles behave according to rules that ensure safe and legally correct behaviour. Since, however, the simulator car is controlled by a human driver, this car may exhibit unexpected and unsafe behaviour.

(c) Scenarios whilst driving in the Small World
Comparability of the results was ascertained between subjects because all subjects encountered the same situations during the experiments. These situations were predetermined by means of a scenario editor. A scenario is a constellation of situations that occur when the simulator car arrives at a certain point in the network of roads. This implies that traffic density and interactions with other traffic were under experimental control.

The scenarios selected for testing of the GIDS system were the following. Note that by "car" we mean the GIDS car in which the subject is driving; by "vehicle" we mean a simulated car in the simulated world.

Scenario 1. The car is driving straight ahead, whilst a leading vehicle decelerates in such a way that it forces the subject to decelerate at 5 m s^{-2}. Scenario 1 was encountered four times during a trip in the simulated world.

Scenario 2. The car is approaching an intersection regulated by traffic lights, while it is closely followed by a vehicle. The traffic light is green. When the subject is near the traffic lights, the lights turn yellow forcing the subject to decelerate by at least 3 m s^{-2} in order to be able to stop in time. A reaction time of 1 second was taken into account. Scenario 2 was encountered once during the trip.

Scenario 3. The car is approaching an obstacle located in the same-lane. In order to pass the obstacle it is necessary to change lanes. A stream of approaching vehicles (from the opposite direction) forces the vehicle to stop. Scenario 3 was designed to study the performance of drivers with and without support in a situation where they were forced to perform a combination of tasks, i.e., assessing the speed of other traffic, acceleration, use of mirror, and steering. Scenario 3 was encountered once during the trip.

Scenario 4. The car is approaching an intersection and intending to turn left when a vehicle approaches from the right. When this vehicle has crossed the intersection and the car is about to turn left, a second vehicle approaches from the opposite direction. Scenario 4 was encountered three times.

Scenario 5. The car is approaching a right-of-way intersection. A vehicle approaches at high speed from the right and, against the rules, takes right of way. Scenario 5 was encountered once during the trip.

Scenario 6. The car is approaching a T-junction (driving on the main road). Just before crossing the right-hand road, the telephone rings. The right-hand road is quite narrow and not easily visible. Scenario 6 was encountered four times during the trip.

Scenario 7. The car is following a curve in the right lane. An obstacle is obstructing the driver's lane, immediately beyond the curve. The car is forced to stop for the vehicle because a stream of vehicles is approaching in the other lane. Scenario 7 was encountered twice during a trip.

These scenarios were considered to be representative of critical situations in which GIDS may provide the driver with relevant support.

Method

The GIDS support conditions

Each subject drove in one of three conditions. In the 'Control' condition subjects received no support other than route guidance. That this was taken as the basic condition was because it was considered self-evident that navigation support by itself would improve navigation performance, so that a comparison with, for instance, a condition in which subjects would have to find their way by means of a map, was superfluous. Route guidance was always given both verbally and on the visual display.

In the 'Non-integrated' condition the support modules were used independently of workload. In the 'Integrated GIDS' condition the support offered by the different modules was scheduled according to workload.

The following Table 10.1 shows which actions do appear in the seven scenarios under consideration.

Table 10.1 Actions of support modules during scenarios

Driver support	Scenario number						
	1	2	3	4	5	6	7
Route guidance message	*	*	*	*	*	*	*
Accelerator feedback	*	*	*	*	*	*	*
Steering wheel feedback	*		*			*	
Suppressed telephone call[*]						*	

[*] The suppressed telephone call is to the integrated support condition

Design of the experiment

The three support conditions were tested between subjects. Three groups of drivers with different levels of an experience/age combination participated in the experiment. The design used thus was a 3x3 between subject design.

Driver groups

Fifty-seven drivers were hired through an advertisement in regional papers. They were differentiated according to sex, age, and driving experience. Thirty-nine drivers were male and 18 were female. Three groups were distinguished: Young Inexperienced drivers ($n=28$), Young Experienced ($n=15$) and Older Experienced ($n=14$). Criteria for the level of experience were number of kilometres driven in the last year, number of kilometres driven in last five years, and time of licensing (Chapter 3, see also Groeger et al.,1990). The range of age in the group labeled as young was 18 to 25 years, and the range of age in the group labeled as older was 40 to 55 years.

Procedure
Instructions given to the subjects before the experimental drive included information on the type of support they could expect and an explanation of the route guidance messages. Before the experimental trip subjects drove for about 10 minutes in the car to familiarize themselves with the simulator and the task. Subjects also received instructions on how to provide workload ratings during the experiment.

Dependent variables
The Subjective Workload Assessment Technique (SWAT) was used to measure workload as experienced by subjects during the experiment. SWAT has three subscales. One pertains to experienced time pressure, one to experienced mental effort, and the third to experienced psychological stress. Ratings are on a 3-point scale, where 1 is 'little' and 3 is 'much'. Ratings are added to obtain an overall SWAT score indexing workload (Biers & McInerney, 1988). SWAT ratings were obtained from the subject after the occurrence of each separate scenario.

The following performance measures were available for statistical analysis:

−   the number of accidents that happened to drivers;
−   the number of navigational errors;
−   the distributions of driving speeds, accelerations, and decelerations;
−   headway to preceding cars in terms of distance and time;
−   gap acceptance (time-to-collision) when overtaking;
−   driver reaction times to events;
−   SWAT ratings, on the three subscales combined;
−   the numbers of (discrete) actions performed by GIDS.

Results

Condition and performance
In scenario 7 there was a main effect of support condition on performance. In this scenario the car negotiates a curve, beyond which it encounters a stationary vehicle. It cannot overtake the vehicle as other vehicles are approaching. Compared with the condition with no assistance subjects in the support conditions had (1) a lower speed when entering the curve; (2) a longer headway in terms of both time and distance; (3) a smaller deceleration and (4) a larger accepted gap size (see Figure 10.1).

No main effects of the support condition on performance measurements were found in the scenarios 1 to 6.

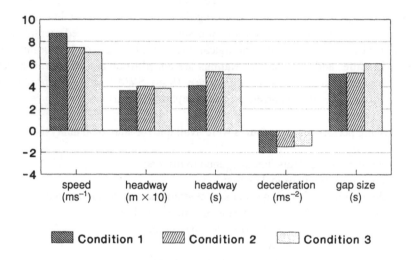

Figure 10.1 Performance in scenario 7 in three conditions (1 = Control, 2 = Non-integrated, 3 = Integrated GIDS)

Occurrence of accidents

Four accidents occurred, one in the no support condition, and three in the non-integrated support condition. No accidents occurred in the integrated GIDS condition. All accidents were caused by the young inexperienced and young experienced subjects.

Route guidance messages

The support on route following was given on a GIDS video screen. Pictograms of the next decision point were presented. An arrow indicated the direction to take. The route-following messages were presented in all conditions. Altogether of 31 errors were made by 23 subjects. Thirty four subjects did not make any error in following the route in the simulator. The proportion of navigation errors was approximately equal for all conditions. In the integrated GIDS condition messages were filtered and scheduled on the basis of estimated workload and assigned levels of urgency. Navigation messages generally received low priority and, in the integrated GIDS condition, were either given earlier or somewhat later in situations with high workload. Findings showed that this did not give rise to significantly more navigation errors. Findings did, however, show a difference in the proportion of navigation errors by the young drivers versus the older drivers: respectively young inexperienced, young experienced, and older experienced (0.60 and 0.61 vs 0.36).

Support via the steering wheel

Feedback about the position of the car was given to the driver via a discrete pulse on the steering wheel. This type of support was implemented in scenarios 1, 3 and 7. The system gave its warnings as follows: in scenario 1 GIDS acted via a pulse on the steering wheel if subjects made an attempt to overtake the decelerating leading vehicle. In scenario 3 a pulse on the steering wheel was provided when subjects initiated, or were engaged in a dangerous overtaking manoeuvre. In scenario 7 (while negotiating a curve) GIDS reacted with a pulse on the steering wheel whenever the driver swerved from the right lane. Immediately after the bend subjects had to stop behind a stationary vehicle whilst a number of other vehicles approached. If subjects began a dangerous overtaking manoeuvre in that situation, GIDS also acted with a pulse on the steering wheel. No conclusions can be drawn from the study about the effect of this type of lane-keeping support on the occurrence of accidents of conflicts. Significant differences that can be attributed to the action of the steering wheel were found in scenario 3. In the support conditions subjects accepted longer gaps before overtaking, that is, in the control condition the accepted gap size was 4.9 seconds, in the non-integrated condition 5.3 seconds and in the integrated GIDS condition it was 5.5 seconds. This same trend was shown in scenario 7 with accepted gap sizes from 5.0 (control), 5.2 (non-integrated support) and 6.0 (integrated GIDS support).

All support initiated by the steering wheel was logged during the experiment. The initiated steering support was dependent on the support condition either given to the subjects, or delayed, or canceled. The condition had an effect on the number of times that steering support was initiated. In the control condition, support via the steering wheel was initiated (but not given) 397 times per subject. In the non-integrated condition and in the integrated GIDS condition support via the steering wheel was initiated 219 and 200 times per subject respectively (see Figure 10.2). This suggests that subjects adapted their behaviour when they received support via the steering wheel, by driving in such a way that the pulse on the steering wheel would occur less frequently.

Support via the accelerator

Accelerator support was available in all scenarios in the support conditions. It consisted of either an increasing or a decreasing counterforce on the accelerator. As mentioned earlier, a main effect of condition was found in scenario 7.

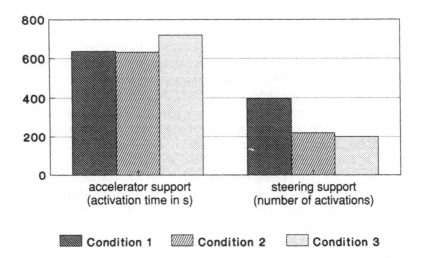

Figure 10.2 The support initiated by the accelerator and by the steering wheel was logged during the experiment. Condition 1 = control, Condition 2 = non-integrated and Condition 3 = integrated GIDS

The action of the GIDS system most prominently present in both support conditions of scenario 7 was an action on the accelerator when subjects approached and entered a bend at excessive speed. Findings revealed a main effect of support condition on the performance measurements speed, headway, and deceleration (see Figure 10.1). These effects can be attributed to the support actions of the intelligent accelerator pedal.

The support initiated by the accelerator was logged during the experiment. The initiated support was, depending on the condition, either presented to the subjects, or delayed, or canceled. In the integrated GIDS condition subjects received more messages via the accelerator than in the two other support conditions (see Figure 10.2). Unlike the responses to the steering wheel messages, subjects did not seem to adapt their behaviour towards a decrease in the number of messages via the accelerator.

Workload
Subjects assessed their workload via SWAT scoring lists. The ratings were quite low, indicating that most subjects had chosen a defensive style of driving in the simulator. Scenario 5 received the highest scores for workload (5.06), whilst scenario 1 got the lowest mean scores (3.52). The SWAT scores showed little difference in the three support conditions.

Interactive communication devices
In the GIDS system optimal use is made of the possibility to present information in modalities other than the visual one. Consequently most messages to the driver were provided via active controls (steering wheel and accelerator pedal), whilst others (e.g., route guidance) were presented verbally and visually. All messages were scheduled in the integrated GIDS condition according to workload. This included the action of the scheduler to delay the telephone ringing just when a subject arrived at a complex junction (scenario 6). The simulation study showed effects of scheduling in scenario 7 where, in the integrated GIDS condition, navigation messages were delayed till the overtaking manoeuvre had been completed. Another finding was that the accepted gap size was largest in the integrated GIDS condition (6.0 seconds compared with 5.1 and 5.2 seconds in the two other conditions).

Discussion
In this simulator study a (limited) set of support messages has been tested, namely those messages that are required for a proper functioning of the system under Small World conditions as defined earlier in the project.

The only scenario where support as offered in the non-integrated and/or the integrated condition of the system did have a direct positive effect on the performance of subjects was in the scenario where drivers had to negotiate a curve, encountering an obstacle in their lane immediately beyond that curve. Other effects were marginal.

We assume that these somewhat unexpected results were due to two factors. First, subjects appeared to drive rather defensively, so that critical situations were less critical than they would have been at higher driving speed. Second, it was observed that in many of the critical situations hardly any scheduling was needed in the integrated GIDS condition, again indicating that circumstances may have been much less critical, in terms of the information offered to the driver simultaneously, than we had anticipated. This assumption should be borne out by creating more complicated scenarios, or by imposing time pressure on the subjects in the simulator.

10.3 Field evaluation

General

The GIDS prototype as installed in the ICACAD vehicle has been evaluated by comparing driving performance in real traffic with and without the prototype.

Difficulties were posed to the evaluation by the fact that some essential GIDS modules were not yet fully operational at the time of testing. Therefore their functions had to be simulated in a form that, to the subjects, would operate in such a way that system performance would, altogether, be indistinguishable from the fully operational GIDS system. These simulations, however, constrained the experimental procedures that were feasible, as will appear from the description of the experimental set-up. This should be understood and appreciated when judging what the evaluation really achieved; see Janssen et al. (1992) for a more extensive description.

Method

The route
The experiment took place in the city of Utrecht (the fourth largest town in The Netherlands with a population of approximately 230,000).The experimental route consisted of three parts ('legs'), each leading to a different destination in the city. The subject's task was to drive to each of these destinations, as specified by the experimenter, in a fixed order.

The route was approximately 9 kilometres long and took about 20 minutes to complete. It was chosen so as to create a variety of situations that would be difficult to handle for someone unfamiliar with the city. Thus, it comprised intersections and roundabouts of often complex geometry, as well as a variety of streets, alleys, and avenues. There were altogether 15 traffic lights along the route. The route also included several pedestrian crossings, and on some parts of the route bicyclists would mix with the motor traffic. Traffic on this route was always busy, but never so busy that the traffic would come to a complete standstill (except at traffic lights).

The GIDS support conditions
Each subject drove in one of three conditions. In the 'Control' condition subjects received no support other than route guidance. In the 'Non-integrated' condition support modules were used independently of workload. In the 'Integrated GIDS' condition the support offered by the different modules was scheduled according to workload.

Route guidance was always given vocally as well as on the visual display. The guidance messages were derived from the predetermined route, whose characteristics (distances between decision points, geometry of decision points) had been programmed into the navigation support module. The module then acted on this preprogrammed knowledge by emitting the appropriate messages at the decision points.

Because no operational sensors to detect lead vehicles were installed at the time, the collision avoidance (CAS) module only acted on the presence of a 'stooge' vehicle. This was an instrumented vehicle, which appeared as a leading vehicle on the route from the second to the third destination. Because subjects in the GIDS vehicle should continue attending to the route-guidance messages considerable care was taken to disguise that the stooge did actually lead the way on that part of the route. Thus, the stooge drove as if it naturally formed part of the traffic stream in which the GIDS vehicle also found itself. To accomplish this, the driver of the stooge had extensively practised procedures for not losing the GIDS vehicle while, at the same time, behaving unobtrusively. Nevertheless, about one driver in five was 'lost' in the experiment either because other vehicles got in between the GIDS vehicle and the stooge, because the vehicles became separated at traffic lights, or because drivers voiced their suspicion that the leading vehicle was actually a stooge.

The required speed and distance data for the CAS module were obtained by on-line calibration of the current positions of the two vehicles by means of active infra-red sensors which were triggered by reflecting material installed at given locations along the roadside (Verwey, Bakker, & Burry, 1992). Information about the stooge's current speed was sent to ICACAD through a wireless data link. This then allowed the ICACAD computer to calculate indicators of headway relative to the stooge such as time-to-collision (TTC). If a 4 s TTC criterion was met, an action of the accelerator followed, that is, an increased counterforce on the pedal.

The stooge vehicle was present during the third (final) part of the route, in all support conditions. In the 'Control' condition, during which the CAS module was not active, the transmitted speed and distance data were only used as measures of car-following performance. When support conditions included the CAS module the accelerator also acted on its own as a speed limiter. The counterforce in that case was added whenever the ICACAD's speed reached the prevailing speed limit. Knowledge of prevailing speed limits was derived from the inventory of the experimental route that had been made in advance in order to obtain the basic data for the navigation module.

In the non-integrated support condition the CAS (plus speed limiter) module was simply added to the basic route guidance module. In the integrated GIDS condition this addition was constrained by workload considerations.

Table 10.2 summarizes how support conditions were assigned to the three different parts of the route. It should be kept in mind that each subject drove the route only once, under one of the support conditions only, because otherwise subjects would gain so much knowledge of the route on their first trial that the trip could not reasonably be repeated under a different support condition.

Table 10.2 Supports and experimental conditions

Part	1	2	3
Route guidance	C/niS/G	C/niS/G	C/niS/G
Speed limiter	niS/G	niS/G	niS/G
Collision avoidance			niS/G
Integration			G

C: Control
niS: non-integrated Support
G: integrated GIDS condition.

Creating an information overload

At two locations an information overload was created in which the GIDS Scheduler, as the essential component of 'Integrated GIDS', had to take action. Thus, the comparison between the 'Non-integrated' and the 'Integrated' support conditions focused on and was restricted to these two occasions.

The locations were on part 3 of the route, where the stooge vehicle was present. The stooge vehicle would brake quite suddenly and forcibly at both locations, whilst the ICACAD driver was engaged in a modestly complex navigational decision. At precisely this moment the telephone would ring, and after two rings a simple question ('When is your birthday?') was asked which the driver had to answer immediately.

In the 'Non-integrated' support condition all this did indeed happen at the same time. In the 'Integrated' GIDS condition, however, the route guidance information was postponed until the driver had resolved the threatening conflict with the stooge vehicle, while the telephone call was postponed even longer, namely, until the navigational information had been fully provided.

It is clear that an information overload of this type cannot be created too frequently during the relatively brief period of one trip (20 minutes). For one, packing the experiment with critical episodes would make it uncharacteristic of real traffic and is therefore likely to induce subjects to adapt their driving style in an equally uncharacteristic (too defensive) manner. Moreover, this would arouse subjects' suspicions as to how they are actually being manipulated, and what role the hitherto 'neutral' leading vehicle might have in this. From an evaluation point of view, however, the fact that one of the essential comparisons of the experiment is based on just a few seconds of data is unfortunate.

Driver groups

Drivers were hired through an advertisement in regional papers. When potential subjects called for information they were first asked about their general knowledge of a number of Dutch cities, including Utrecht, without informing them that the experiment would take place in Utrecht. Only male drivers who said they were hardly or not at all familiar with Utrecht were accepted as subjects.

Drivers were differentiated according to age and driving experience. Three groups were distinguished: Young Inexperienced drivers ($n = 17$), Young Experienced ($n = 14$), and Older Experienced ($n = 21$). Each separate cell of the experimental design contained between five and seven subjects. These numbers apply to those subjects for whom experimental data became ultimately available, that is, they exclude those whose ride had to be terminated for one reason or another or whose data suffered from technical malfunctionings. Criteria for the composition of driver groups were identical to those for the simulation study. The age range of 'young' drivers was from 18 to 38 years, and of 'older' drivers from 40 to 66 years.

Procedure

Instructions were read to the subjects before the start of the experiment. These included an explanation of the type of support subjects could expect. Subjects in the 'Control' condition were told the meaning of what they were going to hear and see. They also received instructions on how to provide SWAT ratings to the experimenter after having completed each separate part of the route (see below). These subjects received no further instruction during the experiment.

Subjects in the GIDS support conditions were also informed about the route-guidance messages and the procedure for obtaining SWAT ratings. Moreover, these subjects were told that the accelerator would act as a speed limiter. At the beginning of part 3 of the route these subjects were informed that the accelerator would, from then on, also function as a CAS with respect to preceding vehicles.

Before the start of the experiment proper, subjects familiarized themselves thoroughly with the vehicle.

The experimental runs took place between 1000h and 1500h, so that there was no risk of being held up in rush-hour traffic. There were always three subjects per day of experimentation. Experiments took place in the month of June 1992.

Apart from the driver there were two passengers in the ICACAD, an experimenter who was also in charge of safety procedures and who could operate the double brake with which the vehicle is supplied, and a technician who sat in the back of the vehicle operating the experimental equipment.

A final aspect of the procedure is that subjects were not actually allowed to make navigation errors. This would have implied allowing subjects to break away from the stooge vehicle. If the experimenter noticed that the subject was beginning to go astray the correct direction was indicated to him and an error was logged.

Dependent variables

The Subjective Workload Assessment Technique (SWAT) was used to measure workload as experienced by the subjects as in the simulator study.

Performance measures taken continuously during driving were navigation errors made, instantaneous driving speed, and instantaneous acceleration/deceleration. Headway to the preceding vehicle – the stooge – was derived from the speed and distance data that were obtained as described in the section on the GIDS support conditions.

After preprocessing of the raw data the following performance measures were available for statistical analysis:

–   the number of navigation errors;
–   the median value of the distribution of driving speeds, truncated at 15 km h^{-1} so as to get rid of very low speeds associated with stops at traffic lights, etc.;
–   the 85th percentile of the distribution of driving speeds, so as to be able to assess possible speed effects in the upper range;
–   the median values of the distributions of instantaneous accelerations and decelerations;
–   the fraction of the total driving time (on part 3 of the route) in which headway to the leading vehicle was below 0.5 s, and the fraction in which it was below 1.0 s;
–   the overall SWAT rating on the three subscales combined. This variable could take values between 3 (very little workload) and 9 (extremely high workload).

Results and discussion

Support conditions and driver age/experience

Analysis of variance (ANOVA) was performed on the preprocessed data. A significance level of 10% was applied because the experiment was not of a strictly controlled nature but had a considerable element of exploration in it. Moreover, because the experiment took place in real traffic the data must be considered as noisy. The following effects were observed:

–   *Navigation errors.* There were no differences between support conditions or between driver groups as to the number of navigation errors made. Navigation errors were in fact a very rare occurrence. Specifically, navigation errors did not occur more often in the 'Integrated' support condition, where navigation was of relatively low priority (at least in the two instances where an information overload had been created).

− *Median speed.* Significant effects on median speed were obtained for the age/experience dimension ($p = 0.07$) and the support dimension ($p = 0.08$). Table 10.3 shows median speeds in relation to age and experience and Table 10.4 shows the effect for the support dimension. Post hoc testing showed that the difference between the two groups of young drivers (that is, the experience dimension) was not significant, but that older drivers were, on average, somewhat slower. By the same test it appeared that there was no significant difference between the non-integrated support and the integrated GIDS condition, but that both differed from the 'Control' condition.

Table 10.3 Effect of age and driving experience on median (truncated) speed (v50) and on 85th percentile speed (v85)

	v50	v85
Young inexperienced	33.8	51.5
Young experienced	34.2	51.2
Older experienced	33.0	50.4

Table 10.4 Effect of support conditions on v50 and v85

	v50	v85
Control	34.2	52.1
Non-integrated support	33.2	50.5
Integrated GIDS	33.5	50.6

− *85th percentile speed.* Significant effects on this variable were as for median speed. The age /experience dimension was significant at $p= 0.10$, the support condition at $p= 0.01$; see Tables 10.3 and 10.4. As with the median speed the effect with respect to driver variables was in age, and not in experience. And with respect to GIDS conditions the effect was in the comparison of the 'Control' condition on the one hand to both the integrated and non-integrated support conditions on the other.

− *Median acceleration and deceleration.* Support conditions were found to have a significant effect on the median value of deceleration ($p= 0.0007$); see Table 10.5. Post hoc testing indicated that the effect was solely due to the 'Control' condition, so that there was no difference between the non-integrated support and the integrated GIDS condition.

Table 10.5 Effect of support conditions on median deceleration

Control	37.3
Non-integrated support	34.8
Integrated GIDS	33.4

– *Headways below 0.5 s and 1.0 s.* Both the percentages of very short and short headways were affected by support conditions. Significance levels were $p = 0.006$ and $p= 0.00001$, respectively. The data are shown in Table 10.6. The effects were both due to the 'Control' condition: there was no significant difference between the non-integrated support and the integrated GIDS condition.

Table 10.6 Effect of support conditions on percentage very short (<0.5s) and short (<1.0s) headways

	% <0.5	% <1.0
Control	3.1	17.3
Non-integrated support	0.6	2.6
Integrated GIDS	0.6	4.6

– *SWAT scores.* Two significant effects were obtained for the overall SWAT ratings as given by the subjects. One was a main effect of support conditions ($p= 0.03$), the other the interaction between support conditions and parts of the route ($p= 0.09$). The extent of these effects may be judged from Table 10.7. The pattern in the data is that a reliable increase already appears on the first parts of the route, where both GIDS conditions function as speed limiters only. However, workload is differentially affected during the critical part 3 of the route. In the 'Non-integrated support' condition workload does increase considerably, relative to the 'Control' condition. In the 'Integrated GIDS' condition this increase is significantly less.

Table 10.7 Effect of support conditions, as an interaction with 'Part of route', on SWAT ratings of experienced workload

	Part		
	1,2	3	Average
Control	3.74	3.67	3.70
Non-integrated support	3.99	4.65	4.32
Integrated GIDS	4.05	4.33	4.19

Discussion

On several parameters of driving performance (speed, level of deceleration, and following headways) it did make a difference whether GIDS support was given, although there are no discernible differential effects of integrated as compared with non-integrated support. However, there are indications that integrating support at certain critical occasions had a beneficial effect on workload compared to non-integrated support. These are, in general terms, the findings of the field evaluation of the GIDS prototype.

GIDS support affected v50 and v85 of the (truncated) distribution of driving speed, as well as the median value of the distribution of instantaneous decelerations. Moreover, the occurrence of short (<1.0s) and very short (<0.5s) headways with respect to a leading vehicle was almost eliminated.

It may therefore be concluded that the prototype has demonstrated, at least in a preliminary way, the usefulness of the GIDS concept in terms of the effects obtained on parameters of driving performance. We have not as yet been able to demonstrate, however, that integrated support is superior to non-integrated support as far as driving performance per se is concerned.

If we now look at the workload as experienced by the drivers, and as indexed by the overall SWAT ratings, we notice that the introduction of non-integrated support is accompanied by a considerable increase in workload. This is in line with our basic expectations of what will happen when there is no coordination of the messages generated by the separate support functions. The increase is of the order of 25 percent (see Table 10.7). Integrated GIDS support succeeds in reducing this by some 8 percent, or approximately one-third of the original increase. This is quite impressive, given that this relative reduction is achieved by confronting drivers only with two cases of information overload. It was only in these two cases that differences between non-integrated and integrated support could possibly have become manifest. What seems not unreasonable to expect, therefore, is that workload ratings will go down still further when more of these critical cases are encountered, that is, when drivers obtain more experience with the support.

It is also remarkable to observe that subjective workload ratings in this experiment were sensitive to support conditions without differential effects of non-integrated versus integrated support appearing in driving performance itself. Apparently, drivers in this experiment were still able to cope behaviourally with the increased workload under non-integrated support. However, the increased ratings in this condition might well be predictive of trouble in case further 'untreated' peaks in information overload were to occur.

Variation in driver age and experience, as introduced in this evaluation, did by itself have effects on behaviour and on experienced workload. Of more relevance, however, is that there were no interactions with support conditions, so that there is no evidence yet that driver age or experience determines how much can be gained from the support. The positive effects of support were equally large for the respective driver groups. We must therefore conclude that we have not been able to demonstrate the differential effects of driver age and experience that we had expected.

10.4 General conclusions: what about GIDS?

The aim of the evaluation studies in the simulator and in the field has been to compare the effects of the different GIDS support conditions, in particular those of integrated versus non-integrated support.

In order to make the relevant comparisons, experiments had to be designed that contained the critical situations that can indeed discriminate between the support conditions. On the other hand it was necessary to resist the temptation to pack the experiments with critical situations only. This would have been likely to cause a general change in our subjects' driving style, yielding results that would surely be unrepresentative of everyday driving.

There are no independent means by which we can check if we succeeded in choosing the optimal experimental settings for making the comparisons of interest. In the simulator study subjects may have been forced into a somewhat more defensive driving style than usual and in the field study there may have been too little criticality in the events that happened.

Given this uncertainty, however, we feel that we have generated some evidence – both in the simulator and in the field study – that GIDS support does indeed make a difference to driving performance and that there exist situations in which there is indeed a differential effect of integrated versus non-integrated support on relevant counts. Evaluation on a much larger scale will, of course, be required to assess long-term effects. Moreover, it will permit us to have another look at the possibility that there are in fact significant interactions between support conditions and driver variables such as age and experience.

A final aspect to be mentioned is that we appear to have established that the GIDS prototype functions well in a strictly technical sense. Throughout the evaluation studies GIDS did what it was supposed to do at the appropriate moments. Of course this is no guarantee that the system will never show failures or breakdowns. In fact, the successor to GIDS, the Ariadne Project will take a very thorough look at its so-called 'system safety', with the aim of avoiding just these types of incidents. Given that the prototype was already reliable enough to support drivers through complex simulated and real environments we anticipate a further successful evaluation (and improvement) of the GIDS system.

References

Biers, D.W., & McInerney, P. (1988). An alternative to measuring subjective workload: Use of SWAT without the card sort. In *Proceedings of the 32nd Annual Meeting of the Human Factors Society* (pp. 1136-1140). Santa Monica, CA: Human Factors Society.

Groeger, J.A., Kuiken, M.J., Grande, G.E., Miltenburg, P.G.M., Brown, I.D., & Rothengatter, J.A. (1990). *Preliminary design specifications for appropriate feedback provision to drivers with differing levels of traffic experience.* Deliverable Report DRIVE V1041/GIDS-ADA 01. Haren, The Netherlands: Traffic Research Centre, University of Groningen.

Janssen, W.H., Kuiken, M.J., Miltenburg, P.G.M., & Verwey, W.B. (1992). *Demonstration and evaluation studies of the GIDS prototype.* Deliverable Report DRIVE V1041/GIDS-MAN 04; GIDS-NAV 05; GIDS-DIS 03. Haren, The Netherlands: Traffic Research Centre, University of Groningen.

Kuiken, M.J., & Miltenburg, P.G.M. (1993). *Evaluation of the GIDS prototype in the small world driving simulator.* Technical Report DRIVE V1041/GIDS-TR6. Haren, The Netherlands: Traffic Research Centre, University of Groningen.

Verwey, W.B., Bakker, P., & Burry, S. (1992). ICACAD: *A tool for human factors evaluation of sophisticated in-vehicle systems in real traffic.* Report IZF 1992 C-14. Soesterberg, The Netherlands: TNO Institute for Perception.

Chapter 11
Impact and acceptance

John A. Groeger, Håkan Alm, Rudi Haller, John A. Michon

11.0 Chapter outline

GIDS has aimed at a definition of standard rules, a protocol for the filtering, prioritization, integration, and presentation of the various sources of information. It also has developed means of supporting drivers' performance of specific aspects of the driving task, with the capability of adapting such support to the needs of individual drivers. In this chapter we attempt to assess the likely acceptance of such new functions in vehicles, and we consider also the impact that widespread acceptance of such technology would have. In Section 11.1, in order to set the background to this discussion of the acceptance and impact of GIDS, we consider some general issues in the marketing of systems such as GIDS, including the aspects which are likely to seem attractive and unattractive to potential users. Obviously the marketing of GIDS will have no small influence on its acceptance by the public, which is discussed in Section 11.2. Acceptance will in turn determine the impact of GIDS, which is considered in Section 11.3. Finally, in Section 11.4, we discuss the collateral impact of the GIDS programme, that is, the extent to which GIDS is capable of generating spin-off not only towards other RTI topics, but also towards technological and behavioural insights of a very general nature.

11.1 Marketing GIDS

The process whereby a new device or product becomes accepted by a large section of the public is complex, and very much the concern of the market researcher. When dealing with the introduction of new technology, particularly where it would conceivably revolutionize some aspect of daily living, physicians, philosophers, and others have a larger role to play (e.g., developments in 'in vitro' fertilization).

Ultimately, however, the public decides. One of the first problems to be solved is therefore how to describe the new product to its potential users in a way that will secure time for its benefits as well as its drawbacks to be fairly considered.

Problems of description and reaction

In describing the product the system designer and developer are facing something of a dilemma. Inevitably, the new system must be described to potential users in terms they are already familiar with (i.e., existing technologies), which may reduce the apparent 'novelty' of the new system. This may in fact make it more 'acceptable' to potential buyers but less worthy of the effort of manufacturing the actual product because it too closely resembles an existing product, unless new benefits can be anticipated. There is a more serious side-effect of describing the system in terms of existing technology, however, in that it encourages the view that only relatively minor changes will be brought about by its introduction, and that 'old values' remain the appropriate yardstick to apply. Objections to the new technology based on a flouting of old values encourage voluble and often emotional reactions which, because they are irrational, prove difficult to dislodge. It may be difficult for end-users to realize that such standards may simply become irrelevant after the widespread adoption of the new technology; thus men waving red flags no longer need to precede automobiles in built-up areas.

On the other hand, the strategy of describing something as revolutionary is one that should be adopted cautiously. At least one section of each community would view the prospect of a revolution unfavourably! Take, for example, humanity's long cherished dream of flight. Except for the more visionary, the prospect of being more than 20,000 metres from the surface of the earth, travelling at the speed of sound, doubtless would have filled man with fear. Offered such a prospect, one imagines that potential customers for such a service would have been few. On the other hand, the chance to visit loved ones on the other side of the globe, to engage in business opportunities and to experience cultures, previously denied because of the time, difficulty, and expense travel would involve, is now very much welcomed. Such changes are not solely the result of clever marketing, but indicate the underlying capacity of living forms to adapt to new circumstances. The age-old fears about the vulnerability of those in flight have not disappeared and they may easily surface again when an aeroplane in flight shudders due to turbulence, or return in a new guise when a new air disaster occurs. The moral of this is that, if "user-acceptance" is to be the touchstone, it is important that potential users be asked the right question; that the unrealistic but nevertheless real fears people may have are anticipated, and that the benefits as well as the drawbacks of any new system receive careful consideration. The difficulty of describing the GIDS system is not resolved, but at least much reduced by the availability of a prototype (see below). This prototype not only demonstrates the feasibility of GIDS, but also offers the clearest insight into what life with GIDS would be like.

Driving with GIDS

By considering what is embodied in the GIDS prototype, as realized in the Small World Simulation and ICACAD, it is possible to convey something of the feel driving with a GIDS-equipped vehicle would have. In the next section we will consider the acceptability of such new in-car functions, but for the moment let us simply describe what driving with a GIDS system would be like for different drivers.

For drivers driving perfectly in an area they know well, driving with GIDS will appear to have little or no overt effect on the way in which the task is performed. There are two important qualifications to this general statement. First, the GIDS system will continue to function under such circumstances as scheduling the presentation of information transmitted to the vehicle so as to prevent overloading the driver, logging environmental and performance information in order to update the on-board driver history, and predicting potential emergency situations which never actually occur because of the driver's appropriate actions. The driver would of course be oblivious to such operations, except for the increased provision of new communication equipment which the GIDS scheduler would now allow to be operated safely. While it is clear that some of what GIDS has to offer would be readily apparent to the potential user, in other cases the advantages of GIDS would have to be inferred from changes in the usability of other in-car facilities.

For the driver driving perfectly in an unknown area, GIDS will bring about an additional change; it will support the driver with navigation information to aid route-finding, without compromising safety. Once again the scheduling and presentation of communications will be determined by the difficulty of the driving task being performed and the ability of the driver to meet its demands.

Finally, a driver who fails to perform the driving task to an appropriate standard will notice considerably more activity on the part of the GIDS system. Depending on how his or her performance deviates from an acceptable standard, this driver will be presented with information indicating that speed (from the intelligent accelerator pedal), heading (from the intelligent steering wheel) or other aspects of aberrant behaviour (from PSALM) require improvement. Such errors will have an additional, observable consequence: they will cause the GIDS scheduler to reduce demands on the driver from non-driving-related tasks (e.g., communications). Driving with GIDS may therefore be safer, easier, and more productive, efficient, economical, or pleasant depending on the desires of the driver and, if the driver wishes, it could attain a higher standard. Of course such advantages are not gained without some concomitant costs being incurred. It is perhaps inevitable that the increased scope for communication with a specific vehicle will infringe on privacy and, to some extent, personal freedom. It is also reasonable to suppose that the capacity of the system to inform the driver that he or she has performed incorrectly could prove extremely irritating. In the next section, we will consider these aspects of the GIDS system as they relate to the overall acceptability of the system.

11.2 Acceptance of GIDS systems

Acceptance by different groups

In discussing the acceptance of GIDS systems, it is important to consider to which actors in the traffic system the technology would seem likely to prove acceptable or unacceptable. Passengers and pedestrians stand to benefit considerably from the widespread adoption of GIDS systems, albeit indirectly. Pedestrians in particular must surely benefit from drivers who are less distracted from the main driving task, be it because the navigation task has been made easier or because distraction from external communication is minimized. Given that GIDS technology, largely through the operation of PSALM, also seeks to improve standards of driving in general, enhanced safety of passengers, pedestrians and cyclists also seems likely. The views of these groups are all too rarely considered in studies of the acceptability of new in-car systems. Effective communication with vehicles allows the traffic manager to re-route or warn drivers in the event of imminent congestion, thus generally smoothing traffic flow, improving environmental conditions, and maximizing the utility of parking and emergency services. While it is not the intention currently to use information available from GIDS for purposes other than supporting the driver of the vehicle to which it is fitted, the release of such information to enforcement agencies or for use in insurance matters would obviously be attractive, but hardly to all parties concerned! If such information were to be used for automatic policing (as was considered in the DRIVE AUTOPOLIS project) or incriminating a driver in an accident after-the-fact (by insurance companies or the police, as was considered in the case of the DRIVE DRACO project on the journey data recorder), drivers would be reluctant to buy cars with such devices. Even if violations of traffic laws detected by GIDS are not translated into actual penalties, drivers may react negatively to being told by GIDS systems that they are exceeding the legal speed limit or entering a bend too fast. Obviously these issues seem likely to make GIDS systems less acceptable to potential users. Reactions to the release of information about drivers, even to the legitimate authorities, and to a lesser extent 'correction' by an electronic system, are frequently emotional and often irrational. Some feel that the use of GIDS information in this way would compromise individual freedom, and might well evoke Orwellian images of *Big Brother*. While one can have some sympathy with this view, such a stance becomes very questionable when the 'individual freedom' which is being protected actually infringes on the freedom or safety of others. Such weighty moral debates and emotional reactions tend to obscure the real issue concerning the electronic collection of information about errors and offences: such systems (assuming their reliability, which is discussed below) do no more than log behaviour as it occurs and reduce the 'chance' element in the detection by enforcement agencies of such behaviours. It is not the electronic system which constrains behaviour, but the legal code enforced by the system. Anticipating an issue raised below, one might add at this point that if people require a reasonable chance of evasion in order to consider a law fair or acceptable, then it says little in favour of the concept of justice enshrined in such a law. Leaving aside this moral debate, it is arguable that, even with the carefully monitored release of information available from GIDS for legal purposes, the benefits of GIDS functions will eventually be seen to outweigh what might now appear to the driver as drawbacks.

Acceptance by individual drivers

Because GIDS seeks to provide information to the driver that is intended to increase safety, and to present or schedule it in a way which also maximizes safety, one of the benefits offered to drivers by GIDS should be increased safety. However, since it has been suggested by many authors (see, for instance, the discussion in Groeger, 1990) that risk information does not always play a determining role in behaviour, it may be optimistic to consider increases in safety as the most important or most 'saleable' aspect of GIDS. Studies of the acceptability of GIDS technology, carried out as part of the evaluation of the GIDS prototype, are particularly relevant here. While 60-70 per cent of those who used the system in the simulated Small World, considered that it would be useful, almost 80 per cent considered that it would enhance safety, and although virtually all drivers said it would be useful for some or all drivers, only about half of the drivers tested said they would use or possibly buy such a system (Kuiken & Miltenburg, 1993). Ironically, an aspect of driver behaviour which is considered to compromise safety actually offers the possibility of improving driver performance although it was not investigated during the system acceptance study. People generally overestimate their driving ability (Groeger and Brown, 1989), considering themselves better drivers than they actually are. This suggests that driving skill is much prized by people. Through its instructional support function GIDS offers the possibility of revealing, and correcting, people's false impression of themselves. Obviously this is something of a two-edged sword, since drivers may not take kindly to constant reminders that they are not as good as they consider themselves to be. However, the facility to customize, or make more private, the operation of PSALM effectively removes the embarrassment or irritation likely to be caused by presenting such messages in the hearing of others. What GIDS offers by way of aiding self-improvement (particularly if accompanied by appropriate inducements) may therefore enhance user acceptance considerably. When the effects of giving drivers feedback on their performance were studied in an evaluation of the PSALM concept, it was found that drivers' self-assessments of their own ability converged on assessments made by a driving instructor, but they remained very different from such an objective assessment of their ability when feedback was not given (see Groeger & Grande, 1992). This study also made clear that how this feedback is given seems likely to affect the driver's reaction to such support, and ultimately the user's acceptance of such a support system. Other features of the GIDS system, for example, the extent to which the system offers improvements in ease of driving, efficiency, and productivity, also seem likely to promote acceptance. Finally, it should be borne in mind that different groups of drivers will require different choice options because of their needs, and that this will also influence acceptance of the GIDS system. Among professional drivers, for example, employers' interests in driver status, road safety, and vehicle location may override some of the objections these drivers themselves may have. People with special needs, or whose mobility is reduced by age or disability, would also have much to gain from the availability of GIDS systems, because these could conceivably be customized to meet particular support needs. The special needs of such groups are in many cases little more than concrete examples of GIDS functionality in general. Indeed, acceptance of GIDS systems by particular drivers or groups of drivers will depend on the degree to which the system meets these drivers' particular needs. A system that uses high technology for its own sake rather than to meet actual needs will eventually not be used. A system that provides support that is inappropriate because it fails to take account of driver capacity (e.g., reaction time to warnings) will end up frustrating the user. Not only must the driver's needs be met, they must also be central to the design. As is clear from the previous chapters dealing with the design and development of the GIDS prototypes, these concerns have been of paramount importance throughout the programme.

An important point to be borne in mind with regard to acceptance of the GIDS system is the issue raised by Bainbridge (1987) in an article on the ironies of automation. Some drivers will resist automated systems because they derive satisfaction from developing and maintaining their driving skills. The resistance of pilots to advice from air-traffic controllers and co-pilots, has shown that this sense of independence and desire for personal control is an important consideration. GIDS has been careful to place the concept of adaptive and instructional support at the core of the system. This affords drivers who care to develop and maintain their skills the opportunity to do so. The fact that such support is based on a personal performance history may help to reduce adverse reactions caused by comparison with external standards of performance. With regard to a reduction in personal control, GIDS at a very early stage took the decision that the ultimate responsibility for all actions would remain with the driver. As a result GIDS has deliberately adopted a 'non-interventionist' type of driving support.

An exception to this principle is low level vehicle control (steering and acceleration). The reason is that in this case the time constants involved in taking appropriate action are really too short to allow warnings or advice by means of displays. Such support has the additional disadvantage in that it is more likely to prove distracting from the task at hand. Nevertheless, even when the intelligent accelerator pedal or intelligent steering wheel operate, ultimate control remains throughout with the driver. The study of user acceptance of the various components of the GIDS system referred to earlier (Chapter 10, and Kuiken & Miltenburg, 1992) shows that personal control is a crucial issue in drivers' reactions to such systems. While about 50 per cent of the drivers questioned considered the intelligent accelerator pedal and intelligent steering wheel useful, about 75 per cent considered these active controls to be unpleasant. This contrasts with the acceptability of an RTI system which does not affect the driver's control. The Navigation function, for instance, was considered both useful and pleasant by well over 95 per cent of those questioned. Finally, and perhaps most importantly, the claims which have been made for the GIDS system, and its acceptance, must fundamentally be moderated by its reliability. Obviously, if the system is to be trusted by drivers who use it, it must perform reliably. If the obstacle detection system is missing important threats or producing too many false alarms, the driver will disregard it. If the navigation system gave directions that could not be followed (such as the advice to turn on to a street that is one-way in the wrong direction) the driver is unlikely to use that system. The evaluation exercises carried out have helped to establish benchmarks for the reliability of the present generation of GIDS systems. These encourage us to believe that systems which are sufficiently reliable to be used under everyday driving conditions will eventually prove possible (see Janssen, Verwey, Kuiken, & Miltenburg, 1992). Unless systems which are at least as reliable as the driver can be developed, drivers are unlikely to wish to use them and legislators will not sanction their use.

11.3 Impact of GIDS

GIDS and RTI

We consider that the GIDS concept constitutes an important, innovative step towards the development, implementation, and acceptance of advanced RTI systems, in particular of those systems that should help road users to cope with the information load to which other RTI applications may be expected to contribute. Given that considerable progress has been made towards achieving the major objectives of the GIDS project, it emerges that GIDS itself, in principle, signifies a breakthrough in RTI. Two features are particularly relevant in this respect. It is easy to conceive of a variety of task domains for offering electronic driver support, each of which would certainly add to the knowledge of a driver about specific driving conditions, and each of which might also contribute to the driver's confusion and workload.

The GIDS project has produced a set of recommendations that will help to avoid confusion and overload. GIDS being a generic concept, these recommendations should be quite independent of specific applications and consequently they should be useful as overall guidelines for the design and architecture of GIDS-compatible applications.

GIDS and safety

It is clearly intended that RTI should offer an important contribution to the safety domain. However, unless the information load that is likely to occur as a consequence of a liberal implementation of RTI is contained in ways that are compatible with the GIDS human-machine interface philosophy, RTI may easily turn out to be a serious threat rather than a relief to traffic safety. An important function of GIDS is to help control the workload of the driver who is facing an increasing amount of information presented by different sources. Even if safety had not had such a high profile within the DRIVE Programme, in the GIDS project a prominent place for safety issues would almost have been inevitable because of the preponderance of behavioural scientists involved (many of whom have extensive records of achievement in the field of industrial and traffic safety). This strength has resulted in the tackling of issues which not only seemed amenable to current technology, but which also offered the potential for substantial safety gains. Thus, for example, driver-reported tendencies to have particular sorts of accidents or to commit certain types of error were used to structure the empirical work carried out and the development of the PSALM instructional support system. GIDS therefore oriented itself towards safety issues from the outset and, as a result, high-risk groups of motorists, particularly young inexperienced drivers, have been a particular focus throughout. A further point related to safety and the researchers involved is that not only have their wider concerns influenced the development of the GIDS system, but working within GIDS will itself exert an influence on other work currently being carried out and planned. Therefore, is seems reasonable to suppose that research outside the current scope of he GIDS project, relating, for example, to driver risk taking, to the particular problems of the ageing driver, or to the special needs of the driver hoping to return to driving after an accident, stroke or serious illness, will yield applications which could easily be incorporated within the GIDS framework. Thus, when considering the impact on safety which is likely to accrue from GIDS, it is also worth considering that the safety benefits which will emerge from the current GIDS programme will be further supplemented by other developments by the same teams.

GIDS and scientific development

GIDS, in aiming at a definition of standard rules, comprising a protocol for the filtering, prioritization, integration and presentation of the various sources of information, has immersed itself in much of what might be regarded as state-of-the-art cognitive science. By attempting to evaluate these rules with an eye on safety, effectiveness, impact on driver behaviour and workload, and (individual) acceptance, the programme has also concerned itself with very practical issues. In the latter stages of the project this has been an important part of the project activities. Much of the evaluation has been based on empirical results of advanced simulator and on the-road experiments (Chapter 10). Even if the evaluation exercises had proved less successful, the building of a reliable prototype of such a complex information processing system would itself have been a considerable achievement. Those of us who have taken part in the GIDS project have welcomed the challenge of working with other disciplines, in spite of the occasional tensions which such different approaches and pressures bring about. Just as in the DRIVE programme as a whole, the exercise of attempting to promote cross-fertilization of disciplines seems likely to prove a continuing success. The ideas that have emerged thus far from the GIDS project will have their own impact on the future of science and enterprise.

11.4 GIDS in a wider context

In the preceding consideration of the factors which seem likely to affect the impact and acceptance of GIDS systems, a number of seemingly extravagant claims have been made about the potential utility of the system. The claims made are indeed considerable, but they are not incautious. Throughout the GIDS project we have endeavoured to remain within the bounds of available (or imminent) road infrastructure, within the bounds of currently available computational devices, and strictly within the bounds of what is known about human performance and cognition. The question might be raised as to why we concentrated on the supportive functions of GIDS, instead of taking a more principled step in the direction of the robot driver (Michon, 1987). The answer is that the introduction of completely automatic (primary control) or advanced co-operative driving (secondary control) requires a traffic environment that is totally different from the present, comparatively unconstrained, environment which admits vehicles, road technology, and drivers, with vastly different performance characteristics The technology push toward automation should not make us close our eyes to the fact that, for a very long time to come, only human drivers will be capable of dealing with incomplete, ambiguous, or even contradictory information, and yet make correct decisions under most, and even very extreme, conditions. Unless major steps towards a much more tightly constrained road environment are taken or, at least, seem within reach, it would be advisable to maintain and even increase the effort devoted to the kind of driver support envisioned in GIDS and to adopt a cautious, conservative attitude towards co-operative or automated driving.

As we have discussed briefly above, the increasing prominence of information technology in everyday life will bring about considerable changes in the way we live. The introduction of RTI systems, as pioneered in DRIVE projects such as GIDS, AUTOPOLIS and DRACO, will change the quality and quantity of information available to those in authority. This should lead not simply to a more stringent application of current law, but to the rigorous reconsideration of any legislation that relates to the traffic system. Thus, as the quality of evidential material improves, traffic law should be brought more into line with the capabilities of drivers and modern vehicles.

It may seem strange not to have addressed the issue of the cost of GIDS technology in a chapter on the impact and acceptance of the GIDS system. This is neither an oversight, nor an underestimation of the importance that commercial pressures will inevitably exert on the further development of GIDS. The GIDS equipped vehicle will undoubtedly be more expensive, but the level of such additional expense and its value for money will depend greatly on developments in technology and road infrastructure over the coming decades. At this point in time, insofar as it is possible to estimate its cost, a GIDS system would prove too expensive and of very limited use to an ordinary motorist, without a huge expenditure on infrastructure to facilitate road-vehicle communication. While the real strength of GIDS lies in its ability to co-ordinate and integrate the operation of different components, some of these, such as route guidance systems and collision-avoidance systems, could be used independently. Thus one can anticipate individual components of an eventual GIDS system being widely available. Obviously, were such component systems to evolve to become standard installations, the additional cost of a full GIDS system would be greatly reduced, perhaps even below the 1,000 ECU threshold which is widely recognized by manufacturers as the price that motorists are likely to pay for an in-vehicle system. The widespread use of such component systems would in all likelihood have another effect: GIDS would become increasingly marketable, since without it the dreaded explosion of uncoordinated information would be at hand. It would then be in the interests of the motorist, the public and, indeed, the manufacturers of the component systems to find a costeffective way of filtering, prioritizing, integrating, and presenting this plethora of new sources of information.

References

Bainbridge, L. (1987). Ironies of automation. In J. Rasmussen, K. Duncan, & J. Leplat (Eds.), *New technology and human error* (pp. 271-286). Chichester, England: Wiley.

Janssen, W.H., Verwey, W.B., Kuiken, M.J., & Miltenburg, P.G.M. (1992). *Demonstration and evaluation studies of the GIDS prototype.* Deliverable Report DRIVE V1041/GIDS-MAN 04/NAV 05/DIS 03. Haren, The Netherlands: Traffic Research Centre, University of Groningen.

Groeger, J.A. (1990). Errors in and out of context. *Ergonomics, 33*, 1423-1430.

Groeger, J.A., & Brown, I.D. (1989). Assessing one's own and others' driving ability: Influences of age, sex and experience. *Accident Analysis and Prevention, 21*, 155-168.

Groeger, J.A., & Grande, G.E. (1992). *Meeting the support requirements of drivers with different levels of traffic experience: An evaluation.* Deliverable Report DRIVE V1041/GIDS-ADA 03. Haren, The Netherlands: Traffic Research Centre, University of Groningen.

Kuiken, M.J, & Miltenburg, P.G.M. (1993). *Evaluation of the GIDS prototype in the Small World simulator.* Technical Report DRIVE V1041/GIDS-TR6. Haren, The Netherlands: Traffic Research Centre, University of Groningen.

Michon, J.A. (2 September, 1987). Twenty-five years of road safety research. *Staatscourant, 168*, 4-6.

Chapter 12
The next steps

John A. Michon

12.0 Chapter outline

In this last chapter we consider the overall achievements of the GIDS Project and
further steps in the development of GIDS. In Section 12.1 a retrospective evaluation of the main GIDS objectives is provided. Section 12.2 then deals with the further evolution of GIDS into a full-fledged RTI system; one concern is the place of
GIDS and other GIDS-like support systems within the future of road traffic and
RTI and another issue concerns the insight gained as a result of the GIDS Project
into the architecture of human and artificial intelligence and into the special cognitive requirements of human-computer systems. Finally brief mention is made in
this section of some points that might be considered as potential spin-offs, that is,
as potential applications of certain results to domains of research and development
that are not strictly concerned with driver support. The chapter ends with some
conclusions and recommendations concerning the GIDS Project as a whole and
with 100 percent hindsight.

12.1 Have the GIDS objectives been achieved?

The GIDS Project has culminated in a working prototype of an intelligent co-driver
system. Altogether the prototype does fulfil most of the promises of the original
GIDS Project proposal. As such the prototype is genuinely more than an approximation of the GIDS concept. It should indeed be appreciated as representing the
first generation of GIDS systems.

   With the completion of the GIDS Project it is now time to reconsider the objectives as specified in the overall project plan (Chapter 1), to see to what extent these
objectives have been met and what further steps may be taken. Also the achieve-

ments should be considered against the backdrop of the overall DRIVE context, including the DRIVE II Programme and the successor of the GIDS Project, the ARIADNE Project. ARIADNE (DRIVE II Project No. V2004) stands for Application of a Real-time Intelligent Aid for Driving and Navigation Enhancement, which sufficiently explains the acronymical reference to the Cretan princess Ariadne, young Theseus' lover.

To repeat, the GIDS objectives were:

(a) Define detailed functional requirements of generic intelligent driver support (GIDS) systems.
This objective has been achieved. The system as it is available now is built on explicit considerations regarding its connections with the outside world, its functional modules, its overall architecture, its human-machine interface requirements, the generic structure for its knowledge base – that is, its representation of the outside world, the driving task, and the characteristics of the driver – and, last but not least, the procedures that operate on this knowledge base.

(b) Determine the impact of the new RTI system on the task representations and behaviours of drivers with respect to planning, manoeuvring, and handling aspects of the driving task.
This objective has been achieved with respect to the domain of action, the Small World, as defined early in the project. Although this by no means exhausts the rich structure of the driving environment or the full spectrum of human behaviour and human error, it demonstrates that driver support as given by GIDS-like adaptive systems does affect the various levels of driver behaviour. Further research is needed to determine the full impact – and the functional limitations – of the GIDS system on driver behaviour in the real world.

(c) Determine the interactive communication (display and dialogue) between the driver and the new RTI system, inclusive of adaptive feedback.
This objective has been achieved with regard to several interfacing modes (video-display, speech generation, push-buttons, keyboard, and tactile feedback through accelerator pedal and steering wheel). Also the objective has been achieved in the sense that a (limited) set of support messages has been tested, namely those messages that are required for a proper functioning of the system under Small World conditions. GIDS is a generic design allowing additional interface components and other messages without the need to change the design of the system.

(d) Develop the required hardware and software that shall lead to the implementation of a prototypical GIDS system. This – limited – prototype shall incorporate a substantive core of the GIDS concept.

This objective has been achieved. The project has delivered an operational GIDS Mark I prototype. Two specimens have been constructed, one having been implemented in an automobile (TNO Institute for Perception's ICACAD vehicle) for real-world demonstration and research. The other is incorporated into the Traffic Research Centre's simulation facility for demonstration, rapid prototyping and research. Both specimens are functionally identical and both display a substantive core of the GIDS concept, in terms of architecture, functional performance, and range of operation.

(e) Determine the impact of systems that meet the GIDS specifications on driving safety, efficiency, training, and system acceptance.
The objective has been met in principle. A limited number of performance tests with ICACAD in real-world conditions and in the Small World Simulator under laboratory conditions, have been carried out. The full range of possibilities remains to be charted and explored. The ARIADNE Project, dealing with the various aspects of the GIDS system's full impact and its relation to other RTI systems is currently in progress in the framework of the DRIVE II Programme. It should be acknowledged that, to a large extent, the success of the GIDS project actually lies in its demonstrating the potential for further research and development, rather than in the outcome of the limited testing programme that was part of the GIDS Project reported in this volume. This *potential* is precisely what conceptually constitutes its added value over and above the proof that the GIDS concept is feasible.

(f) Demonstrate the validity of the GIDS concept in field tests.
This objective has been achieved. The operational performance of the GIDS Mark I prototypes proves the point. It should be added that in achieving this ultimate objective the GIDS Project as a whole has been a success. Despite some organizational difficulties due to the coming and going of some partners, the work has progressed according to schedule and none of the deviations from plan has forced the consortium to change the above objectives in an essential way. In some instances the consortium has chosen to steer a somewhat narrower course than it may have envisioned prior to the project, but the full concept of the GIDS system (as specified in Chapters 1 and 5) has been kept in focus throughout.

These objectives seem to converge on one additional issue. It has been pointed out before that the rapid increase in the number and variety of signals informing the driver of some state of affairs inside the vehicle or in the outside world necessitates the development of consistent protocols for an 'information refinery' such as GIDS. Two types of protocol are needed:

– information exchange protocols for modeling interactive behaviour in driving, to be applied in construction and advice regarding various RTI applications;

such applications, when developed according to these protocols, would be compatible with the GIDS architecture;

— technical protocols for implementing sensors and dedicated applications into GIDS systems; such applications, when realized according to these protocols, would be compatible with the GIDS architecture and, thereby, with each other as well.

These protocols are crucially important for a smooth and efficient introduction of user-oriented RTI systems that can adapt to the demands of the traffic system as it may prevail in the various countries or regions in Europe and elsewhere. With the internal frontiers within the European Community lifted in 1993 it will remain highly desirable to tune the traffic system as much as possible to national or local circumstances without loss of generality for Community regulations. The development of these communication protocols is an extremely complicated affair, however, and the GIDS Project has made only a very modest step in the direction of an operational standard for such a protocol.

12.2 The further evolution of GIDS

Technical evolution

Because GIDS as a conceptual system is generic, the specific details of its implementation do not, in the first instance, matter a great deal. It should be clear, however, that implementing GIDS is an important issue *sui generis*. At this early stage in its development which is definitely opening up new territory in the domain of RTI, the actual demonstration that a working GIDS system can indeed be built is not entirely trivial, as we have found out in the course of the project. With the existence of proof delivered, however, it will be necessary to spend considerable additional effort in further developing the GIDS technology.

The first objective should be to reduce and optimize system hardware and software so as to bring series production of the GIDS system within reach. At present two working prototypes are available, each sufficiently bulky to make them quite impractical for general use. The next step should be to construct a small series of GIDS units for further research, including field trials under realistic driving circumstances. Such units should not exceed the size of a single desktop computer.

Second, as a next step all the software for the GIDS Mark II should be redesigned so as to ascertain a smoother and more efficient interaction between the functional components.

Third, once it is clear that GIDS can be fruitfully extended to support drivers in average urban traffic conditions, and that the impact and acceptance of GIDS are satisfactory, the next step would be to implement GIDS in such a way that it can then be made commercially available. This raises the question of the future of GIDS within the framework of RTI, which is currently marked by a trend towards full automatization. What will ultimately be the lifespan of the GIDS concept, that is, the idea of non-intervening driving support? What is the purpose of further developing of GIDS if the first automatic highway lies just around the corner? These and similar questions can be answered by pointing out that any kind of automation will require a formalization of the traffic environment, the driving task, and the needs, capabilities, and limitations of the driver. This is where the further development of GIDS is certain to break new ground.

Functional evolution

We turn now to the consideration of how GIDS functions should be developed from their present state. It should be pointed out that the evolution of GIDS will proceed, functionally speaking, along two lines: (a) interactions with advanced roadside information sources, or so-called 'intelligent roads' and (b) interactions with on-board information sources or so-called 'intelligent vehicles.'

Further research should extend the range of GIDS Mark I beyond that explored in the present project, because constraints had to be deliberately imposed on the system for practical and sometimes theoretical reasons. An important overarching item on this agenda is to refine the user model and instructional capabilities of the system (the PSALM module).

GIDS as it stands is a conceptually sound, but functionally still restricted prototype of an intelligent driver support interface. The results clearly demand further research and development before the GIDS technology can be brought to the RTI market. The principal objectives for a further development of GIDS functions can be summarized as follows:

(a) Situations
The first objective is to extend the range of events and manoeuvres over which the GIDS system will be able to provide support to drivers. The present range is limited by events and behaviours that were defined for the Small World paradigm. The next step should be to include many more situations that prevail in normal driving, especially including those that characterize average conditions in built-up areas.

(b) User variables
The second objective is to extend the range and number of person-dependent variables which determine driver support. In the present project such variables were restricted to age and experience. The ultimate aim is to provide a better, more reliable, and more user-friendly interface between the human driver and the various

components of the road-traffic system. It is therefore essential to extend the scope and power of the PSALM and Calibrator components.

(c) Human-machine interface

Another next step is to develop further the interfacing protocol between GIDS and RTI systems that provide information about road and traffic conditions, vehicle status, and – if possible – driver status. It is evident that several of the activities to be undertaken within the framework of programmes such as DRIVE II and PRO-METHEUS are supportive of this objective. It is therefore necessary that GIDS be put to the test in several of the contexts that are generated within the DRIVE II context.

Related to this point is the 'added value' of the GIDS Project for our insight into complex human behaviour in connection with (artificial) knowledge-based systems. In this respect the spin-off from the GIDS Project for basic research in cognitive science and artificial intelligence is considerable. As an example, one of the most valuable aspects is that GIDS operates in a complex environment in a real time fashion. Insights and recommendations for the further study of real time interactive intelligent architectures may therefore be distilled from the GIDS experience.

(d) Extended field studies

The fourth objective is to subject one or more Mark II GIDS equipped vehicles to extensive testing of system performance, with respect to systems safety, human-machine interfacing, efficiency and reliability of operation under a variety of realistic conditions and during a sufficiently long period.

Research and development

The objectives specified in the preceding section have some implications for the R&D environment in which future GIDS activities should be embedded. Thus, the technical evolution will require a strong input from the RTI-oriented industry, in cooperation with car manufacturers. This applies specifically to the required hardware development. Whilst software development may proceed fairly straightforwardly along the lines followed in the present GIDS Project, the integration and miniaturization of the system's hardware should be a special concern during the first stage of the ARIADNE Project and it must necessarily precede the field studies that are required to meet the other three objectives specified in the preceding section.

The next steps towards the functional evolution of GIDS in general and the human-machine interface in particular should be specifically directed at extending the range and scope of the GIDS system. As such they must be aimed at the formal dynamic description of situations and events, tasks (driver actions), and driver-related data. This implies that further behavioural studies will be needed in order to acquire the necessary data on which the required process descriptions can be based. These will require a suitable, realistic environment in which the necessary data can be obtained. The nature of the studies thus far is such that a 'built-up' (urban or suburban) environment would be most suitable for the purposes of a GIDS II project; such an environment would seem to be a plausible extension of the Small World to which GIDS has been confined until now.

Finally, substantial progress will eventually require the availability of a small fleet of perhaps two or three dozen GIDS-equipped vehicles. This makes it possible to perform the required comparative field studies in which systems safety, human-machine interfacing, and operational reliability can be tested prior to a further step towards the commercial production of a GIDS system.

These steps cannot all be made at the same time. But this should not be seen as an insurmountable problem, since it will remain of the essence of the GIDS system to evolve stepwise and in a modular fashion.

It is possible, as well as acceptable, that GIDS be sooner or later incorporated in one or several quite specific RTI projects. It can profitably become part of, for instance, projects that deal independently with navigation, active roadside beacons, route guidance, parking management, or traffic monitoring. In each of these cases the information needs and workload of the driver require study under more or less realistic operating conditions. This will provide the opportunity to answer a large number of specific GIDS-related questions at the same time. In other words, the generic character of GIDS should continue to make it a very flexible and adaptive system to study in a meaningful way under a variety of conditions.

The results in wider perspective

Apart from these project-related achievements, a number of more general aspects and features of complex skilled performance and adaptive support appear to be open to further exploration and, eventually, perhaps exploitation as a result of the GIDS Project. These aspects and features include both technology and knowledge and are specified in the following points. It should be emphasized that these points are listed because they appear promising and not because any partner in the GIDS Consortium was, or is, necessarily committed to actually exploiting one or more of them. In other words, the following list should be considered only as a series of generalized recommendations for further research and exploration.

–   Basic knowledge about the performance of various categories of drivers who are operating under a variety of circumstances, specifically including their performance when using GIDS. This knowledge might be implemented in a query base, accessible for R&D purposes in the RTI domain, legislation and enforcement, vehicle design, etc.;

- A formal (algorithmic and perhaps partly heuristic) description of the driving task which is extendible to other complex behaviours, to be applied in the construction of expert systems and adaptive monitoring systems, both in traffic and elsewhere;
- Basic insight into knowledge acquisition by trainee and novice drivers, and in the role of instructional feedback to be applied in adaptive tutorial systems (driver training programmes);
- Principles of real-time dialogue management and real-time action scheduling for use in interactive and adaptive human-machine systems, both in traffic and in other task environments;
- Rapid prototyping technique (simulation) for testing and evaluating various aspects of a task environment or support procedures;
- Special purpose GIDS systems for categories of road users with atypical characteristics or specific task requirements, including the elderly, the handicapped, and professional drivers. Each of these would require a special task analysis geared to the idiosyncrasies of the target group.

12.3 Epilogue

How far have we come with GIDS from an outsider's point of view?

An outsider would be likely to conclude that GIDS was a very ambitious project and that it resulted in a substantial amount of work of a very high technical quality. He or she might even add that the GIDS concept may be ahead of its time.

It may be argued, however, and not without reason, that the overall impact of the GIDS system on traffic safety must still be determined and this will indeed require extensive additional testing. It is easy to agree on this issue. First the field trials have been less comprehensive than we had anticipated. But gradually a second point became clear: no system of the complexity of GIDS can be tested to satisfaction in both a simulated and a real environment within the few months that were available. Currently, however, a considerable programme of tests is being carried out as an integral part of the DRIVE II ARIADNE Project (Project V 2004) that should eventually provide additional insight into the system's potential impact on road traffic safety. This should be considered a satisfactory impact on the DRIVE II Programme.

Given the limited programme of field and simulator trials carried out thus far, can we come to a preliminary conclusion that GIDS does indeed what it is supposed to do, namely reduce the driver's workload?

Adding RTI functions, such as navigation or anticollision information to the driving task may lead to a considerable increase in driver workload. This confirms one of the basic suspicions that motivated the GIDS Project in the first place: the threat of information overload as a result of new RTI is apparently real. In those circumstances in which these extra RTI functions do indeed lead to a sharp increase in workload, the introduction of the integrated GIDS support function – the quintessential GIDS: filtering, scheduling, and presentation – does lead to a significant reduction in workload, relative to the load imposed by the non-integrated ensemble of RTI applications (Chapter 10). Under the prevailing experimental conditions, this reduction was not quite large enough to overcome the extra load caused by the extra RTI functions but this is not so worrying after all. Had we been able to study drivers after they had been driving with an active GIDS system for some time, a different result would probably have been obtained. However, prior to the experiment all drivers were totally unfamiliar with the kind of driver support provided by GIDS and they received only very limited advance training.

Importantly, the positive effect of the support provided by GIDS emerged exactly under circumstances where such an effect is to be expected most, that is, at a point where the driver was suddenly and unexpectedly required to brake forcefully, to make a navigation decision, and to answer a call on the car telephone at the same time. This is where filtering, scheduling, and presentation matter most. In other, less complex driving situations no significant differences were observed between the unintegrated and the integrated (GIDS) support mode. Under the latter circumstances it was found, as a rule, that in the both the unintegrated and the integrated mode less workload was imposed on the driver than in the condition without support (see Chapter 10).

In summary, whilst acknowledging once again the need for extensive field trials, the GIDS system seems to be doing precisely what one would expect it to do: *GIDS does reduce workload under circumstances where information might otherwise exceed acceptable limits.*

The PSALM module was not completed and, as a result, could not be tested. This has been a cause for dissatisfaction among the GIDS Consortium. However, it was not omitted as a matter of principle but largely, if not entirely, as a consequence of the delays incurred by the withdrawal of an important industrial partner at a most inconvenient point in the project's development. PSALM is now being implemented and will be an integral part of the GIDS Mark II under construction in the ARIADNE Project.

The success of the Small World paradigm is difficult to overstate. Not only is the Small World of considerable value for behavioural tests, and for rapid prototyping and benchmark testing of GIDS functions (as argued in Chapter 9), but it has already attracted worldwide attention as part of a new generation of (intelligent) driving simulators. As an important next step, following naturally from its present status, research is now under way to further develop this new simulator technology in the context of virtual reality graphic simulation.

From an outsider's perspective it might seem that the relevance and the transferability to real-life driving situations of the Small World methodological framework was not demonstrated in the GIDS Project. Such a conclusion would reflect a general concern about the comparability of data obtained in real driving situations and those obtained in simulator studies more than anything specifically attributable to the GIDS Project. There is evidence – though not from our own test trials – that the relation between the two contexts is consistent over a fairly wide range of driving tasks. This augurs positively for the outcome of future studies that seek to integrate information obtained under real-world and Small World driving conditions. More importantly, however, in the field trials carried out thus far the Small World framework *was* in fact tested on the road, thus allowing an analysis of relevance and transferability.

No cost-benefit analysis has been carried out. The need for such an analysis is evident, but it will be appropriate only at a later stage in the development of the GIDS system. To try one now would be a rather vacuous exercise. In fact, several unsuccessful attempts at a preliminary cost-benefit analysis have been made. If anything these efforts seem to suggest that the GIDS system might become cost-effective for certain categories of professional drivers (e.g., delivery vans and maintenance vehicles) and for handicapped drivers. There are, however, too many uncertainties to justify a proper discussion of these attempts, at least for the time being. A proper cost-benefit analysis will be part and parcel of the DRIVE II ARIADNE Project now in progress.

For an interested outsider the documentation of the GIDS Project must be a delight. At last here is a project whose every step can be traced in detail! That it is difficult for the reader to integrate all of the 34 deliverables resulting from the project, painstakingly produced by ten workteams composed of scientists and engineers in 13 separate locations and in five linguistic domains, in the course of a four year period, should be evident. Only on the basis of the present volume will it be possible to judge the degree of conceptual and technical coherence achieved towards the final stages of the GIDS Project. In addition, the size of this comprehensive output also reflects the involvement of relatively many behavioural and cognitive scientists. For them, more than for engineers, extensive technical reports and journal publications are part of the communicative conventions of their discipline. Importantly, and certainly not in the last place, the documentation properly reflects the complexity of the intellectual interchange between the GIDS Project and mainstream psychology and computer science.

Finally it may seem to an outsider that the GIDS Project could not acquire, nor develop much of its required technology. This observation invites two comments.

First, the basic feature of GIDS is, once again, its generic nature. This genericity allows its range of application to be extended smoothly whenever additional RTI systems become operational. With this proviso most of the technology that is *intrinsically* required for a functioning GIDS system *has indeed been developed* in the course of the project although, admittedly, its level of (intelligent) functioning can be substantially improved. As a matter of fact it will be improved in the course of the ARIADNE Project.

Second, at the start of the DRIVE II Programme there has been little, if any, impact of GIDS on the shape of that programme. This may be a consequence of the extreme push towards on-site implementation that governed much of the preparatory phase of DRIVE II. Perhaps this initial lack of impact will soon turn out to be an advantage. Incorporating a generic but innovative and not yet finalized design such as GIDS into a heavily implementation-oriented programme such as DRIVE II could well put a premature brake on the development of GIDS, thus introducing a severe risk of the system's mental retardation. Moreover, one of the basic issues is how to connect all the RTI applications produced by the DRIVE II Programme to the driver. GIDS should provide a solution to these difficulties, but currently the creators of GIDS feel that they would be vulnerable to severe, and justified, criticism had they promised to provide such a solution now, ready for immediate use and providing direction to other developments under DRIVE II. That would have been counter-intuitive in the light of the 12 years it normally takes an engineering development to reach a marketable level of maturity.

In conclusion then – speaking for the investigators who have been involved in the development of GIDS – we feel confident that the project has made a valuable contribution towards the development of intelligent driver support, and more generally towards the understanding and, hence, the formal description of driver behaviour. There is little doubt that the achievement will have its impact on further developments of RTI technology for years to come. Its performance as a real-time system offering support in a complex real-world task, puts GIDS in line with some very advanced work in robot intelligence – the extra revenue being, in our opinion, its sophisticated real-time human-computer interface.

Author index

Aasman, J., 20, 40, 95
Abelson, R.P., 40, 92
Adams, B.B., 21, 23, 39, 78, 92, 95-97
Adams, C.C., Jr., 61
Akyürek, A., 92
Allen, J.F., 92
Allen, R.W., 63
Allsop, R.E., 22
Alm, H., 33-52, 53-66, 113-146, 120, 217-227
Avolio, B.J., 41

Bainbridge, L., 222
Baker-Ward, L., 54
Bakker, P., 207
Ben-Ishai, R., 41
Berggrund, U., 41
Berlin, M., 116, 117
Biers, D.W., 201
Biesta, P.W., 45
Blaauw, G.J., 45
Bobrow, D.G., 29
Boden, M.A., 92
Boff, K.R., 71
Bovy, P.H.L., 115
Bragg, B.W.E., 34
Braune, R., 44
Brendemühl, D., 35
Brookhuis, K.A., 35
Brouwer, W.H., 36
Brown, I.D., 22, 34, 38, 41, 44, 130, 132, 155, 176, 200, 221
Brown, J.S., 14
Burger, W., 27
Burry, S., 147-173, 207

Card, S.K., 83
Cavallo, V., 23
Cohen, R., 92
Curry, G.A., 45

Davis, J.R., 116, 117
De Velde Harsenhorst, J.J., 95

De Waard, D., 131
DeBald, D.P., 61
Dellen, R.G., 29, 35
Detweiler, M., 45
DRIVE, 4
Duncan, J., 38

Eggemeier, F.T., 44
Engels, K., 29, 35
Evans, L., 35, 38, 121

Färber, Br.A., 29, 124, 163
Färber, B., 19-32, 124-126, 128, 163
Feinstein, M.H., 54
Fenton, R.E., 124
Ferrell, W.R., 64, 123
Finn, P., 34
Flach, J.M., 124
Fontaine, H., 35
Foulke, E., 64
Frazer, D., 36

Gaillard, A.W.K., 45
Garfield, J.L., 54
Gibson, J.J., 23
GIDS, 4 8
Gilson, R.D., 124
Godthelp, J., 19-32, 41, 124-126, 163
Grande, G.E., 130, 132, 155, 176, 200, 222
Groeger, J.A., 19-32, 34-36, 38, 113-146, 155, 176, 200, 217-227

Haller, R., 217-227
Handler, J., 92
Harper, J., 39
Hart, S.G., 44
Hennings, U., 61
Hess, R.A., 124
Hicks, T.G., 45
Hieatt, D.J., 45
Hills, B.L., 41
Hosman, R.J.A.W., 124

ISO, 28

Jagacinski, R.J., 124
Janssen, W.H., 45, 53-66, 113-146, 195-215, 223
Järmark, S., 61
John, B.E., 83
Johnson, L., 41

Kahneman, D., 41, 44
Kaufman, L., 71
Kelley, C.R., 124, 138
Kelly, J., 89
Klein, R.H., 63
Koning, G.J., 115
Kramer, A.F., 44
Kroeck, K.G., 41
Kuiken, M.J., 33-52, 113-146, 155, 176, 195-215,
    221, 223

Labiale, G., 19-32
Laurent, M., 23
Lester, J.F., 35
Lockwood, C.R., 35
Lotan, M., 41
Lourens, P.F., 95

MacDonald, W.A., 44
Matthews, M.L., 34
Maycock, G., 35
McCormick, E.J., 123
McDermott, T.T., 93
McInerney, P., 201
McKnight, A.J., 21, 23, 39, 78, 92, 95-97
McLoughlin, H.B., 89-112
McRuer, D.T., 63
Meister, D., 21
Michon, J.A., xiii-xvii, 3-18, 20, 33-66, 69-87,
    89-112, 158, 160, 217-239
Miltenburg, P.G.M., 38, 130-132, 155, 176, 196,
    198, 200, 206, 221, 223
Modjtahedzadeh, A., 124
Moran, A.R., 34
Moran, T.P., 83

Naab, K., 128
Näätänen, R., 34
Neumann, O., 43, 45
Newell, A., 83
Nilsson, G., 41
Nilsson, L., 36, 61, 120
Norman, D.A., 29
Noy, Y.I., 44

O'Hanlon, J.F., 35

Panek, P.E., 41
Panik, F., 124
Parker, P.M., 38
Pelz, D.C., 36
Perrault, C.R., 92
Piersma, E.H., 69-87, 109, 147-173
Popp, M., 29
Posner, M.I., 54
PRO-GEN, 28, 29

Queen, J., 27
Quenault, S.W., 38

Reid, G.B., 44
Rich, E., 93
Rissland, E.L., 54
Robertson, S.A., 22
Rockwell, T.H., 61, 116
Roscoe, A.H., 44
Rosenbaum, D.A., 54
Rothengatter, J.A., 33-52, 130, 132, 155, 176, 200
Rühmann, H., 64, 123
Rule, R.G., 124
Rumar, K., 41

Sabey, B.E., 41
Sanders, A.F., 43 44
Sanders, M.S., 123
Savelid, J., 61
Schank, R.C., 40, 92
Schenk, N., 35
Schiff, W., 64
Schmandt, C.M., 116, 117
Schmidt, U., 35
Schneider, W., 45
Schraagen, J.M.C., 113-146
Schuman, S.H., 36
Schumann, J., 113-146, 163
Sheridan, T.B., 64, 123
Sirevaag, E.J., 44
Slack, G., 27
Sleeman, D., 14
Smiley, A., 3-18, 35, 40, 53-66, 69-87, 93, 94, 158
Stadler, M.A., 44
Staughton, G.C., 41
Staveland, L.E., 44
Stillings, N.A., 54
Streeter, L.A., 61, 116
Summala, H., 34
Svenson, O., 34

Tate, A., 92
Theeuwes, J., 41
Thomas, J.P., 71
Toland, C., 93 98
Treat, J.R., 41

Van der Mey, P., 131
Van der Vaart, J.C., 124
Van Kampen, L.T.B., 36
Van Winsum, W., 89-112, 113-173, 175-191
Van Wolffelaar, P.C., 35, 36, 175-191
Verwey, W.B., 33-52, 61, 69-87, 113-173, 196, 206,
    207, 223
Vitello, D., 61, 116

Warnes, A.M., 36
Wasielewski, P., 121
Webster, E., 69-87, 89-112, 160
Weir, D.H., 63
Weister, S.E., 54
Welford, A.T., 34
Wetherell, A., 45
Wickens, C.D., 43, 44, 123
Wierwille, W.W., 45
Wilde, G.J.S., 45
Williams, P., 38
Wonsiewicz, S.A., 61, 116
Wontorra, H., 113-146

Yeh, Y.Y., 44

Zwahlen, H.T., 61

Subject index

Acceptance of the GIDS system, 217-228
  age, 222
  disability, 222
  enforcement, 220
  existing technology, 218
  instructional support, 221-222
  law, 221
  legal code, 221
  marketing, 217-220
  navigation system, 223
  overestimation of driving ability, 221
  personal freedom, 219
  Personalized Support And Learning
    Module, 221
  presentation of information, 219
  privacy, 219
  user –, 190, 218
  violations of traffic laws, 220
Accidents, 20, 22, 130, 196
  involvement in, 35-36, 41, 46, 130, 189
  near –, 130
  occurrence of, 202
Active control(s), 120, 123-129, 163, 223 (see
  also GIDS functions)
  (active) accelerator pedal, 120, 123-125,
    163, 168, 203-204, 223
  Car Body Interface, 150
  closed-loop control, 125, 127-128
  critical curve driving, 126-127
  driver support, 125
  driver-car interface, 125
  feedback, 126-128
  headway support,124
  lane-changing performance, 125
  lateral control, 126, 129
  longitudinal control, 129
  open-loop control, 125, 128
  proprioceptive-tactile feedback, 124
  speed, 79
  steering, 79
  (active) steering wheel, 124, 163, 168, 203,
    223

stimulus-response compatibility, 123
  time-to-line-crossing, 126
  time-to-collision, 127
  torque shift, 124, 128
  visual flow, 126
  why-a-warning function, 163
  workload, 123
Actual driver model, 59
Adaptation, 74-75
Adaptive control, 55
Adaptive support, 129-136 (see also GIDS
  functions)
  augmenting, 129
  driving instructor, 132, 134
  driving lessons, 132
  driver training, 132-133
  feedback about performance, 135, 230
  oral support, 132
  overtaking manoeuvres, 134
  performance of drivers, 134
  performance profiles, 135
  Personalized Support And Learning
    Module, 134-135, 151, 155-156, 158
  'tutoring' potential, 135
Adaptivity, 84-86
Age, 35-36, 80, 197, 209-210, 213, 222
Aggression, 34
Alcohol usage, 34
Analyst module, 60
Analyst/Planner, 82-83, 93, 105, 109-110, 151,
  155-156, 186
  deviant actions, 83
  remedial actions, 93
  Manoeuvring and Control Support Module,
    149-150, 153, 155, 158
  sensors, 109
  Workload Estimator, 46, 85, 150, 155, 158
Anti-collision, 79 (see also Collision avoidance)
Applications, see GIDS functions
Architecture, 105-109 (see also
  Analyst/Planner; Dialogue Controller;
  PSALM)

Architecture (continued)

    knowledge of the driving task, 105
    knowledge of the driving situation, 105
    pseudo-code, 106
ARIADNE project, 136, 151, 156, 230-231,
    236-238
    'system safety', 214
Artificial intelligence, 54, 89, 91, 234
    user modelling, 91
Augmentation, 54
Automatic action, 74
Automation
    technology push, 225

Behaviour
    dangerous, 188-189
    knowledge-based, 40, 47
    risky, 34
    rule-based, 40, 47
    skill-based, 40, 47

Car Body Interface, 150
Car following, 79, 123-124, 207
CAS, see Collision avoidance system
Cognitive architectures, 8
Cognitive modelling, 92
    precognitive loop, 91
    rule-based representation, 92
Cognitive science, 54, 225, 234
Collateral
    activities, 79
    behaviour 65
Collision, 63
Collision avoidance support, 119-123, 207 (see
    also GIDS functions)
Collision avoidance system, 62, 119-120, 148
    accelerator pedal, 120, 123
    activation, 119
    active control, 120
    'brick wall' criterion, 119
    car following, 123, 207
    collision configuration, 119
    discriminative power, 62
    evasive actions, 120
    side effects of, 121-123
    'stooge' vehicle, 207, 209
    time-to-collision criterion, 119, 121, 123
Communication
    protocol, 79, 149
    structure, 78
Computational theory, 92

Constraints, 75-80
Control, 63-64, 74
    active, 120, 123-129, 149 (see also Acitive
        control(s))
    closed-loop, 125, 127-128
    lateral, 63-64, 126, 129
    longitudinal, 63-64, 129
    manual, 142
    open-loop, 125, 128
    speech, 142
    voice, 162
Control level, 29
Control support, 163-168
Controls, 27-28
Cost-benefit analyssis, 238
Critical manoeuvres, 72
Critical situation(s), 82
Curve driving, 126-127

Dangerous behaviour, 188-189
Design considerations,69-88
    constraints on prototype, 86
Design modularity, 71
Designer, 70-71
Detectable driver actions, 98
Detectable driver task
    categorization of, 98-100
Development, 234-235
Dialogue controller, 60-61, 64, 82-84, 105,
    109-110, 135, 151, 155-163
    allocation patterns, 84
    critical path analysis, 84
    I/O devices, 157
    nonconflicting interfacing, 84
    resource allocation, 83
    time allocation, 83
    workload adaptivity, 157-158
    workload estimation algorithm, 109
    workload indicators, 109
Dialogue Generator, 150, 158-159, 187
Dialogue management, 83
DRIVE, 4, 8
DRIVE II, 230, 231 (see also ARIADNE)
Driver, 6
    actions, 85, 100
    behaviour, 84, 92
    characteristics, 12, 33, 58
    errors, 130
    failure, 20
    experience, 73, 136
    intention, 4, 13
    information needs, 40-42

mental models, 117
motivation, 28, 41
needs, 70-71, 117
performance, 72, 83, 125, 134
plans, 40
reaction times, 74
safety, 91
scenarios, 40
scripts, 40
state, 35
support, 53-68, 72-73, 75, 125, 136, 222
support systems, 4, 6, 10, 56, 56-61, 80
training, 21, 132-133
workload, 42-46, 74
Driver model, 95
driver behaviour, 84
look-up tables, 84
neural networks, 84
precognitive loop, 91
production systems, 84
Driver modelling, 91-94
deviant behaviour, 93
Driver tasks, 190
Driver task modelling, 39
Driver training programmes, 236
Driving
ability, 41
environment, 62, 75
experience, 37-38, 197, 209
instructor, 132
lessons, 132
load, 45
overestimating ability, 221
performance, 131, 213
situations, 101-105
skills, 20
style, 38, 120, 214
Driving simulator, 148, 168, 183 (see also
Simulator; Small World simulation/simulator;
TRC-simulator)
Driving task, 11, 19-20, 29, 34, 56, 58, 62-63,
72, 78, 90, 105, 110, 136, 157, 238
analysis, 21-24, 39-40
demands, 74
knowledge of, 91
support, 59
Drug usage, 34
Dynamic control modelling, 92
precognitive loop, 91
rule-based representation, 92
Ecological optics, 23
Environment-vehicle-driver-system, 19-20

Environmental conditions, 11
Ergonomics, 27, 70
Estimating workload (see also Workload estimator)
dynamic adaptivity, 84
Ethernet, 168, 170
European Community, 232
Evaluation studies, 195-216, 225
acceleration, 211
accelerator support, 203
car following, 207
critical episodes, 208
deceleration, 211
driving performance, 213
experimental route, 206
field evaluation, 206-213
field experimentation, 196
field-studies, 214, 234-235
field-tests, 231
headway, 211
information overload, 208
integrated GIDS support condition, 196-197,
204-208, 211-213, 237
intelligent accelerator pedal, 204
non-integrated support condition, 196-197,
205-208, 211-213
performance measures, 201
performance of the GIDS system, 198
route guidance messages, 202-203, 207
scenarios, 198-203
simulator study, 198, 205, 214
Small World simulation, 198
steering wheel support, 203
Subjective Workload Assessment Technique,
201, 205, 209-210, 212-213
support condition(s), 201, 206-210, 212-213
workload, 197, 200, 203, 205, 210, 213
Evolution of GIDS, 232-236
functional, 233-234
technical, 232-233
Experience, 36-39, 73, 80, 210, 213
Experienced drivers, 45, 91, 130, 197, 209
Experiments (see also evaluation studies)
real-world, 191
simulator, 191
Expert system, 65

Feedback, 63, 65, 126-128
about performance, 135
behavioural, 64-65
instructional, 74-75
self-explanatory, 28
Field study, see Evaluation studies

Genericity, 5, 8
Generic Intelligent Driver Support (GIDS), see
    GIDS
GIDS
    application(s), 85, 140 (see also GIDS functions)
    bus-architecture, 81
    co-driver, 64
    co-driver support, 86
    communication protocol, 79
    concept, 69, 230-231
    cost-effective, 238
    defensive driving style, 214
    definition, 3-4
    design, 69-75
    design specifications, 7
    functional characteristics of, 12-14
    functional complexity, 78
    functional and operational requirements of, 7
    functions in, 113 (see also GIDS functions)
    hardware, 81, 163
    knowledge base, 73
    messages, 82, 163
    modules, 149
    objective(s), 6-8, 229-232
    origin of the project, 8
    project, 8, 10, 225, 229, 232, 236-237
    prototype, 7, 17, 23, 78-79, 86, 109, 135,
        147-149, 151-152, 156-158, 162-163, 168,
        172, 178, 189, 196, 221-222, 230, 232
    support, 47, 212, 214 (see also Evaluation
        studies)
    support function, 40
    support messages, 148
    support system, 23
    system, 26, 33, 35, 38, 40, 55-56, 65, 69, 71-76,
        78, 81, 83, 91, 94, 96, 98, 100, 105, 110,
        124,136, 140, 147, 149, 218, 222, 224-226,
        230,237
    system characteristics, 73
    system evaluation, see Evaluation studies
    task representation, 85
    testing programme, 231
GIDS architecture, 11, 13, 105-109, 147-174, 232
    (see also Active control(s); Architecture;
    Hardware)
    Car Body Interface, 150
    communication protocol, 149
    Control support, 163
    Dialogue Generator, 150, 158-159, 187
    hardware engineering, 148
    implementation, 148
    ISA-bus, 163

    Manoeuvring and Control Support Module,
        149-150, 158
    Navigation module, 149
    Scheduler, 149-150, 158-159, 190
    sensors, 109, 151-153
    software engineering, 148
    traverser module, 179-181
GIDS calibrator, 136-139, 234
    alerting messages, 138
    appropriate level of support, 137
    control functions, 138
    critical setting, 139
    individual preferences, 136-139
    long-term adatation, 137
    menu-driven 139
    safety features, 137
    short-term adaptation, 137
    user preferences, 157
    vital messages, 138
    warning messages, 138
    workload, 137
GIDS functions, 113-146
    adaptive support, 129-136
    active control, 123-129
    application, 140
    benchmark testing, 237
    calibrator, 136-139 (see also GIDS calibrator)
    cognitive resources, 141
    collision avoidance support, 119-123
    dynamic scheduling, 140
    intelligent support system, 114
    interaction clusters, 140, 142
    manual control, 142
    motor resources, 141
    navigation support, 113-118
    priority messages, 142
    rapid prototyping, 231, 236-237
    speech control, 142
GIDS intelligence, 89-112, 148, 186
    information processing load, 90
    knowledge, 89-90
    processing resource
    reasoning, 89
'Good' driving, 57
Grundy system, 93
    attribute value pairs, 93

Handicapped drivers, 236
Hardware, 81, 163
    audio channels, 168
    car telephone, 167
    dashboard keypad, 163

Ethernet, 168, 170
LCD-CGA screen, 163
miniaturization of, 234
speech recognizer, 164
speech sample editor, 164
Human information processing, 54
Human performance, 71
Human-computer systems, 229
Human-machine interactions, 83
Human-machine interface, 79-80, 224, 234
Human-machine interfacing, 12

ICACAD, see Instrumented CAr for
Computer-Assisted Driving
Impact of the GIDS system, 217-228
safety, 224
workload, 223
Individual differences, 23
Individual preferences, see GIDS calibrator
Inexperienced drivers, 45, 130, 209 (see also
Novice drivers)
Information
auditory, 116
exchange protocol(s), 72, 231
explanatory, 79
overload, 208, 236
presentation of, 26, 219
processing, 34, 42
refinery, 231
remedial, 75
tutorial, 79
visual, 116
Information-induced errors, 41-42
Instruction, 64-65
Instructional feedback, 74-75, 221
Instrumented CAr for Computer-Assisted
Driving, 15, 148, 151, 153-154, 164-165,
168-171, 206-207, 209, 219, 231
Instrumented vehicle, 131
Intelligence, 5, 89, 91, 93, 149, 183-184
artificial, 54, 89, 91
Intelligent
architectures, 85, 92
co-driver system, 229
driver support (systems), 46, 56-61
roads, 233
tutoring systems, 14
vehicles, 177, 233
Interactive communication, 230
Interface(s), 71, 83
components, 79
driver-car, 125

menu-driven, 139
soft key, 161
user, 161
Integrated manoeuvres, 83
ISA-bus, 163

Knowledge-based systems, 234

Landmarks, see Navigation
Lane changing, 96-97
performance, 125

Manoeuvring, 25, 62-63, 74
collision avoidance, 62-63
level, 29
task levels, 44
Manoeuvring and Control Support Module,
149-150, 153, 155, 158 (see also
Analyst/Planner)
MAU, see Multi-attribute utility
MCSM, see Manoeuvring and Control
Support Module
Mental
dispositions, 80
load, 59
model(s), 62, 73
workload, 26, 43, 61
workload assessment, 45
Messages, 82-83, 135, 163
alerting, 138
allocation, 83
priorities, 60, 142
priority value, 84
support, 148, 159, 205
structure, 83
vital, 138
warning, 138
Model Human Processor, 83
Modular architecture, 71
Modularity, 71
Modularization, 149
Modules, 149
communication protocol, 149
interactive traffic simulator module, 179
Navigation, 149
support, 200
traverser, 179
Motivation, 34
Multi-attribute utility, 114
Multiple resource theory, 43
Multitasking, see Collateral behaviour

Navigation, 25, 61-62, 73, 79, 153-155
    errors, 118
    landmarks, 117-118
    level, 29
    map-based, 61
    module, 149
    route choice, 61
    route guidance, 79
    system(s), 61-62, 109, 110, 114, 117, 151
    task, 73
    task levels, 44
Navigation support, 113-118 (see also GIDS
    functions)
    auditory information, 116
    distance criteria, 114
    intelligent support system, 114
    left/right instructions, 117
    oral messages, 116, 118
    route choice criteria, 114
    route guidance messages, 202-203, 207
    route maps, 116
    route preference, 115
    route selection, 114-115
    route selection algorithm, 114, 118
    system acceptance, 114
    time savings, 114
    visual information, 116
    visual messages, 118
Neural networks, 84
Novice drivers, 91, 155, 197

Obstacle(s), 62
    detection, 60-61

Perception-action compatibility theory, 64
Perceptual enhancement, 54
Performance
    criteria, 58, 64
    profiles, 64, 135
    support, 21
Person-related factors, 80
Personalized Support And Learning Module,
    134-135, 151, 155-156, 158, 221, 224,
    233-234, 237 (see also Adaptive support)
    criterion of abnormal performance, 135
    self-improvement, 221
Physical dispositions, 80
Planning, 73
Primary task performance, 43
Production system concept, 92
Professional drivers, 236
PROMETHEUS, 10, 234

PSALM, see Personalized Support And
    Learning Module

Rapid prototyping, 231, 236-237
Real-time
    dialogue, 236
    to operate under – conditions, 85
Recommendations, 229, 235
Reference driver, 56, 58
Reference driver model, 14, 24, 56-57, 93
Remedial information, 75
Research, 234-235
Resource allocation, 83
Risk, 34
Risk taking, 224
Road
    categorization, 24
    delineation, 25
    environments, 11
    infrastructure, 225
    network, 151, 181-182
    signs, 24
Road Transport Informatics, 4, 8, 20, 28, 29, 36, 56,
    86, 217, 223-224, 226, 230, 231, 234-236, 238
    user-oriented, 232
Robot driver, 8, 225
Route (see also Navigation, Navigation Support)
    choice, 61
    guidance, 79, 202-203, 207
    preference, 115
    selection, 114-115
RTI, see Road Transport Informatics

Safety, 99, 197, 224
Save driving
    motivation, 80
    personality, 80
    sequencing of actions, 98
Scheduler, 149-150, 158-159, 190
    complex timing, 190
    situations of differing complexity, 190
Scheduling, 83
Secondray task technique, 43
Selective attention, 41
Sensors, 109, 151-153, 182-183
Sensory Integrator, 186
Simulator, 175
    experiments, 177
    graphical, 175, 178-179
Situation analysis, 101-105
    rule base, 101-105
Situation descriptions, 105-109

rule-base, 101-105 (see also Analyst/Planner)
Skill, 34
Small World, 11, 72, 76-77, 82-83, 95-96,
101-105, 110, 131, 157, 172, 230, 237
paradigm, 40, 237
range of acceptable values, 82
scenarios, 131, 190, 198-203
topography, 76
Small World simulation/simulator, 15-17, 72,
77, 148, 151, 175-192, 219, 221, 231 (see
also Simulator)
car cabin, 179
car initializer utilities, 185
decision rules, 183
experimental control, 190, 191
experiment, 191
functional architecture, 184
interactive traffic simulator module, 179
intelligence, 183-184, 186
'intelligent' vehicles, 177
locomotor sensations, 176
logical structure, 181-182
model of human driving, 183
network editor utilities, 185
road network, 181-182
scenario editor, 190, 198
scenario editor utilities, 185
sensor information, 177
Sensory Integrator, 186
simulated scenery, 180
standard traffic regulations, 177
structure of, 177-181
study, 198, 205, 214
test environment, 186-190
traffic environment (dynamic/static), 175,
181, 184, 187-190
utilities, 184-186
vehicle dynamics model, 179-180
virtual 'drivers', 184
Smart
cars, 10
roads, 10
Speech output, 167 (see also Hardware)
Standard traffic regulations, 177
'Stooge' vehicle, 207, 209
Sub-analyst, 60
Subjective Workload Assessment Technique,
201, 205, 209-210, 212-213 (see also
Evaluation studies)
Support, 6
condition(s), 201, 206-210, 212-213
functions, 11, 55-56, 78-79 (see also GIDS)

messages, 148, 159, 205
performance, 21
requirements, 78
Support system, 23, 56-61, 74, 80 (see also
Driver support systems)
'tutoring' potential, 135
SWAT, see Subjective Workload Assessment
Technique
System acceptance, 114
System design, 69-88
System evaluation, see Evaluation studies
Systems
co-driver, 4
collision avoidance, 62
driver support, 4, 6, 10, 56-61
environment-vehicle-driver, 19
human-computer, 229
intelligent tutoring, 14
knowledge-based, 234
navigation, 61-62, 109, 110, 151
obstacle detection, 61
tutorial, 236

Task analysis, 23, 39, 94-100
categorization of detectable driver tasks, 98-100
detectable driver actions, 98
task descriptions, 97
task organization, 96, 98
temporal ordering, 94, 98
Task representation, 85
Telephone, 162, 167
Time allocation, 83
Time-and-motion study, 83
Time-to-collision, see Collision avoidance
system; Active control(s)
Time-to-line-crossing, see Active control(s)
TLC, see Time-to-line-crossing
Traffic environment (dynamic/static), see Small
World simulation/simulator
Traffic law, 226
Traverser, 179-181
TRC-Simulator, 152-153, 168 (see also
Simulator; Small World simulation/simulator)
accelerator pedal, 168
driving simulator, 168
GIDS prototype, 168
steering wheel, 168
TTC, see Time-to-collision
Tutoring
intelligent systems, 14, 236
potential of, 135

Urban traffic, 232
User acceptance, 190, 218
User interface, 160-161
    soft key, 161
User Modelling, 91 (see also Grundy system)
    beliefs, 92
    defiant behaviour, 93
    illocutionary acts, 92
    Persona model, 93
User needs, see Driver needs
User preferences, 157
User-friendliness, 70
Utilities, 184-186
    car initializer utilities, 185
    network editor utilities, 185
    scenario editor utilities, 185

Vehicle control, 25
    task levels, 44
Visibility, 25
Voice control, 162-163
    why-a-warning function, 163
Warning
    message, 75
    signals, 74
Workload, 27, 42, 62, 74, 80, 84-85, 90, 123,
    137, 140, 142, 157, 159, 197, 200, 203,
    205, 210, 213, 223, 237
    adaptivity, 157
    assessment, 43-46 (see also Subjective
        Workload Assessment Technique)
    constraints, 159
    estimation algorithm, 109
    estimator, 46, 85, 150, 155, 158 (see also
        Estimating workload; Analyst/Planner)
    indicators, 109
    mental, 26, 43, 61
    physiological measures, 44
    subjective estimates, 44

Young drivers
    accident involvement, 41

Milton Keynes UK
Ingram Content Group UK Ltd.
UKHW031532071024
449327UK00005B/106